# The E-MYTH
# Contractor

# The E-MYTH Contractor

WHY MOST CONTRACTORS'
BUSINESSES DON'T WORK
AND WHAT TO DO ABOUT IT

# MICHAEL E. GERBER

**HarperBusiness**
*A Division of* HarperCollins*Publishers*

HarperCollins books may be purchased for educational, business, or sales promotional use. For information please write: Special Markets Department, HarperCollins Publishers Inc., 10 East 53rd Street, New York, NY 10022.

FIRST EDITION

DESIGNED BY MARY AUSTIN SPEAKER

Library of Congress Cataloging-in-Publication Data

Gerber, Michael E.

    The e-myth contractor : why most contractors' businesses don't work and what to do about it / Michael E. Gerber.

    p. cm.

ISBN: 0-06-621468-8

1. Construction industry—Management.  2. Contractors.  3. Cash flow.  I. Title.

HD9715.A2 G456 2002

692'.8'0681—dc21                      2001039688

02   03   04   05   06   RRD/   10  9  8  7  6  5  4  3  2  1

To my children, from the youngest to the oldest,
Alex, Sam, Hillary, Kim, and Shana—my wish is to be
the best father in the world, the reality, somewhat
different . . . but through it all know how deeply
I love you all. To your clarity, your loving natures,
and your liberation.

When we express our true nature, we are human beings.

*ZEN MIND, BEGINNER'S MIND*
SHUNRYU SUZUKI

The basic difference between an ordinary man
and a warrior is that a warrior takes everything
as a challenge while an ordinary man
takes everything either as a blessing or a curse.

DON JUAN IN *TALES OF POWER*
CARLOS CASTANEDA

# CONTENTS

# FOREWORD

I am not a Contractor, though I have helped dozens of Contractors reinvent their businesses over the past 30 years.

I like to think of myself as a thinker. Yes, I like to do things. But before I jump in, I prefer to think through what I'm going to do and figure out the best way to do it. Over the years, I've made it my business to study how things work and how people work. Specifically, how things and people work *best* together to produce the best results.

The result has been a series of books I've authored, called *The E-Myth,* and I've also built my own company—E-Myth Worldwide—which, for the past 25 years, has helped thousands of small business owners, including many Contractors, reinvent the way they do business by applying The E-Myth principles to their business. That is what this book is about: how to produce the best results as a real-world Contractor.

This book, intentionally small, is about big ideas. It offers both an introduction to my other E-Myth books and insights into the thinking that is the basis for my business, E-Myth Worldwide, as well as the books.

My aim with this book is to help interested Contractors begin the exciting process of completely transforming the way they do business. In fact, Contractors who have read it say that *The E-Myth Contractor* has been the most important book on business and contracting they have ever read! I hope you will find it to be as helpful.

A few words about what you will find in *The E-Myth Contractor.* Unlike other books on the subject of the contracting business, I don't try to tell you how to do the work you do. Rather, I share with you some profound insights into how great businesspeople think, whether they are Contractors or not.

I'm convinced that this is a new perspective on the kind of thinking that Contractors must adopt if their businesses are to flourish. I call it Strategic Thinking, as opposed to Tactical Thinking.

In Strategic Thinking, also called Systems Thinking, you, the Contractor, will begin to think about your *entire* business—the broad scope of it—instead of focusing on its individual parts. You will begin to see the End Game, instead of being consumed as most Contractors are in the day-to-day routine of your business—the work I've come to call "doing it, doing it, doing it."

Strategic Thinking will enable you to create a business that works *apart* from you instead of *because* of you. Strategic Work will enable you to begin to understand the profound difference between going to work *on* your business and going to work *in* your business.

Once you've read Chapter 1, you may read this book in any order you want. The topics included here are the very issues all Contractors face daily in their businesses. You know what they are: People, Money, Management, Sub-Contractors, and many more.

You will note that my approach to these subjects is unique. For it is my view that until you approach these subjects and your business Strategically, you will never own a business that's friendly to you, that's kind to you, that takes care of you and your family.

Instead, if you approach these subjects the way virtually all Contractors do—that is, Tactically—you will be consumed by your business, frustrated with people, frantically searching for money, uncertain what to expect, and thus always surprised when you face a potential catastrophe, a job that goes horribly wrong, or an important person who leaves your employ for no apparent reason while you pick up the pieces.

In short, *The E-Myth Contractor* is my attempt to help you restart your business and turn it into the kind of business it should be—a business of pure joy!

So I hope you enjoy this small book. I'm certain if you follow its lead, it will take you to places you never thought you'd go. And if *The E-Myth Contractor* proves as worthwhile to you as I believe it will, all I ask in return is that you tell other Contractors about it. Better yet, get them a copy of their own. Then it will be *your* gift as well as mine.

## ABOUT THE E-MYTH

Let me add a few more words here about The E-Myth. The E-Myth is the entrepreneurial myth. It is the subject of my first four books: *The E-Myth, The Power Point, The E-Myth Revisited,* and *The E-Myth Manager.*

This book, *The E-Myth Contractor,* is my way of introducing you to The E-Myth Point of View, which has enabled tens of thousands of people all over the world to transform the way they do business.

The E-Myth says that most businesses fail to fulfill their potential because most people starting their own business are *not* entrepreneurs at all, but rather what I call *technicians suffering from an entrepreneurial seizure.*

After all, most electrical Contractors were electricians, most plumbing Contractors were plumbers, and most framing Contractors were framers. And just because you're a great electrician doesn't mean you know how to run an electrical contracting business.

The E-Myth says that when technicians suffering from an entrepreneurial seizure start a business of their own, they will almost always end up working themselves into a frenzy, doing just about every job without a break—14 hour days, 7 days a week. Going from one job to the next, one project to the next, one chore to the next, often without the time to think.

In short, The E-Myth says that most Contractors don't own a true business—most own a job . . . doing it, doing it, doing it . . . hoping like hell to get some time off, but never figuring out how to get their business to run without them.

And if your business doesn't run well without you, what happens when you can't be in two places at once? Ultimately it fails. And it's happening throughout the world even as you're reading these words. The tragic fact is that fully 80 percent of all Contractors' businesses will fail, most in the first 5 years. Most of the rest will fail in the next 5 years. And they all fail for the very same reason: The owner of the business doesn't know how to build a business that works! Yes, it's true. No matter how much a Contractor knows about the work of plumbing, electricity, or framing, none of that expertise is in itself sufficient to build a successful business. It's only the beginning.

The good news is that it doesn't have to happen to you. The E-Myth philosophy I am about to share with you in this book has been successfully applied to tens of thousands of businesses just like yours with extraordinary results.

It's my thesis that The E-Myth not only *can* work for you, but that it *will* work for you. And in the process it will give you an entirely new experience of both your business and your life.

To your future and your life. Good reading.

Michael E. Gerber
Petaluma, California
June 2001

# ACKNOWLEDGMENTS

To Steve Levee, my friend and the creator of The Ethical Home Story in this book, your love, commitment, and amazingly prolific ability to tell every joke you've ever heard are a constant inspiration to me. Thanks for being my buddy;

To every contractor I have ever worked with, without whom this book would never have been written;

To all our Certified E-Myth Consultants worldwide, if this doesn't help you I don't know what will. Thanks for your commitment to transforming small business throughout the world;

To Steve Boga, whose dogmatic approach to getting this book done has been both exasperating and liberating, thanks for doing what you said you were going to do. It's not only appreciated, but respected as well;

To all the folks at E-Myth Worldwide, past, present, and future, whose determination to bring the dream back to small businesses wherever we've found them is not only inspiring but remarkable, thank you, thank you, thank you, you know who you are!

To Tom Bardeen, whose Contractor clients will attest to the power of The E-Myth and Tom Bardeen's personal, lifelong, unparalleled commitment to it and to them, thank you for your heart, your passion, and your integrity;

And, finally, to Jeff Gospe, my friend and the president of E-Myth Worldwide, you know what you've committed and you know how much I respect you; thank you from the bottom of my heart for "pulling the pieces together."

# INTRODUCTION

Someone once said, "Be careful what you wish for because you just might get it." Most Contractors I've met can appreciate this thought. Because at some point in their career, they have asked themselves, "Why in the world did I become a Contractor?"

And while I don't know how you've answered that question in the past, I am confident that once you understand the Strategic Thinking laid out in this book, you will answer it differently in the future.

However, if the ideas here are going to be of value to you, it's critical that you begin to look at yourself in a different, more productive way.

I am suggesting you get beyond identifying yourself primarily as a person who knows how to do the work of your business, who is an expert at the technical stuff, who has always stepped into the fray and gotten the work done, no matter what the cost.

Instead, I'm going to propose that you forget about the daily grind and begin to think about what a business really is, how a contracting business—or, for that matter, any business—*really* works, as opposed to the way you have come to *believe* it works based upon your experience.

In this book I'm going to take you back to the very beginning of your business, to again think of it as a *product* rather than as a place to go to work; as an *invention* rather than as a place where people work; as a *vision* rather than as employment for you and others.

That is what I propose to do for you in this book: to spur your critical thinking about the *business* of Contracting, rather than the *work* of Contracting. Because it is my deeply held belief—based upon the success we have had helping thousands of owners and managers— that it is only when you begin to think about the business of Contracting and the exciting creation and conduct of that business, that you will begin to discover the secret underlying the creation of every stunningly original business.

The E-Myth says that only by conducting the business of Contracting in a truly innovative and independent way will a Contractor ever realize the unmatched joy that comes from creating a truly independent business, a business that works *without* you rather than *because* of you.

The E-Myth says that it is only by learning the difference between the work of a *business* and the business of *work* that Contractors will be freed from the predictable and often overwhelming tyranny of the unprofitable, unproductive routine that consumes them on a daily basis.

The E-Myth says that what will make the ultimate

difference between the success or failure of your Contracting business is first and foremost how you *think* about your business, as opposed to how hard you work in it.

So, let's think it through together. Let's think about those things—Work, People, Money, Time— that dominate the world of Contractors everywhere.

About Planning. About Growth. About Management. About Getting a Life!

Let's think about improving you and your family's life through the development of an extraordinary business. About getting a life that's *yours*.

# CHAPTER ONE

# The Story of Richard and Anne

I didn't have to invent lies—my tongue did it
by itself, and I was often astonished at how
clever and farsighted a tongue can be.

"A FRIEND OF KAFKA"
ISAAC BASHEVIS SINGER

Every business is a family business. To ignore this truth is to court disaster.

This is true whether or not family members actually work in the business. Whatever their relationship with the business, every member of a Contractor's family will be greatly affected by the decisions a Contractor makes about the business.

Unfortunately, unless some family members are actively involved in the business, many Contractors tend to compartmentalize their lives, seeing their business as separate from their family. These Contractors see their business as a job, and therefore none of the family's business.

"This doesn't concern you," says the Contractor to his wife.

"I leave business at the office and my family at home," says the Contractor, with blind conviction.

And I say with equal conviction: "Not true!"

In actuality, your family and business are inextricably linked to each other. Believe it or not, what's happening in your business is also happening at home.

Consider the following, and ask yourself if each is true:

- If you're angry at the business, you're also angry at home.

- If you're out of control in your business, you're equally out of control at home.

- If you're having trouble with money in your business, you're also having trouble with money at home.

- If you have communication problems in your business, you're also having communication problems at home.

- If you don't trust in your business, you don't trust at home.

- If you're secretive in your business, you're equally secretive at home.

And you're paying a huge price for it!

The truth is that your business and your family are

one—and you're the link. Or you should be. Because if you try to keep your business and your family apart, if your business and your family are strangers, you will effectively create two worlds. Two worlds that can never wholeheartedly serve each other. Two worlds that split each other apart.

Let me tell you the story of Richard and Anne.

Richard and Anne were married, with two children. They lived near Sacramento, California. They dearly loved each other, were active members of their church, participated in community organizations, and spent "quality time" together. All in all, they considered themselves one of the most fortunate families they knew.

Richard had worked as a framer for 8 years while diligently studying at nights for his Contractor's license. When he finally had his license in hand, he started his own home-building firm.

Before making the decision, he and Anne spent many nights talking about the move. Was it something they could afford? Did Richard really have the skills necessary to make the business a success? Was there enough business to go around? What impact would such a move have on their lifestyle, on the children, on their relationship? They asked all the questions they needed to answer before going into business for themselves.

Finally, tired of talking and confident that he could handle whatever he might face in business, Richard committed to starting his own home-building company. Because she loved Richard and did not want to stand in his way, Anne went along, offering her own commitment to help in any way she could.

That's how Best Construction, Inc. got its start. Richard took out a second mortgage on their home, quit his job, and set up shop in their garage.

In the beginning, it went well. A building boom had hit Sacramento, and Richard had no trouble getting framing sub-contracts from the hard-pressed builders he knew in the area. His business expanded, quickly outgrowing his garage.

Within a year, Best Construction employed four full-time framers. It also employed a bookkeeper named Robert to take care of the money. A young woman named Sarah handled the telephone and administrative responsibilities. Everyone worked out of a small office in a strip mall in the middle of town. Richard was ecstatic with the progress his young business had made.

Of course, managing a business was more complicated and time-consuming than working as a framer. Richard not only supervised all the jobs his people did, but he was continually looking for work to keep everybody busy. In addition, he did the estimating, collected money, went to the bank, and waded through illimitable piles of paperwork. Richard also found himself spending more and more time on the telephone, mostly dealing with customer complaints and nurturing relationships.

As the months went by and the building boom continued, Richard had to spend more and more time just to keep things rolling, just to keep his head above water.

By the end of its second year, Best Construction employed 12 full-time and 8 part-time people, and had moved to a larger office downtown. The demands on Richard's time had grown with the business.

He began leaving home earlier in the morning, returning home later at night. He rarely saw his children anymore. But Richard, for the most part, was resigned to the problem. He saw the hard work as essential to building the "sweat equity" he had long heard about.

Money was also becoming a problem for Richard. Although the business was growing like crazy, money always seemed scarce when it was really needed. He had quickly discovered that Contractors were often slow to pay.

As a framer, he had been paid every week; as a Sub-Contractor, he often had to wait—sometimes for months. Richard was still owed money on jobs he had completed more than 90 days before.

When he complained to late-paying Contractors, it fell on deaf ears. They would shrug, smile, and promise to do their best, adding, "But you know how business is."

If Richard pressed the point, the Contractor would say, "Look, I'll write the check when I can. If that isn't good enough, I'll just have to get someone else to do the job." Afraid of losing a contract, Richard would back off.

Of course, regardless of whether Richard got paid, he still had to pay *his* people. This became a relentless problem. Richard often felt like a juggler dancing on a tightrope. A fire burned in his stomach day and night.

Making it worse, Richard began to feel that Anne

was insensitive to his troubles. Not that he often talked to his wife about the company. "Business is business" was Richard's mantra. "It's my responsibility to handle the company and Anne's responsibility to take care of the children, the house, and me."

Anne's seeming lack of concern rankled Richard. Didn't she understand that he had a business to take care of? That he was doing it all for his family? Apparently not.

As time went on, Richard became more consumed by Best Construction. Not surprisingly, Anne grew more frustrated by her husband's lack of communication and lack of interest in her and the children. She persisted in quizzing him about what was going on in the business, why he always looked so stressed. She pressed him to spend more time with his family.

Robert, the bookkeeper, was also becoming a problem for Richard. Robert never seemed to have the financial information Richard needed to make decisions about payroll, purchasing, and the general operating expenses of the business, let alone how much money was available for Richard and Anne's living expenses.

When questioned, Robert would shift his gaze to his feet and say, "Listen, Richard, I've got a lot more to do around here than you can imagine. It'll take a little more time. Just don't press me, okay?"

Overwhelmed by his own work, Richard usually backed off. The last thing Richard wanted was to upset Robert and have to do the books himself. He could also empathize with what Robert was going through, given the company's growth over the past year.

Late at night in his office, Richard would sometimes recall his days as a framer. He missed the simple life he and his family used to have together. Then, as quickly as the thoughts came, they would vanish. He had work to do and no time for daydreaming. "A business is a great thing," he would remind himself. "I simply have to apply myself, as I did in school, and get on with the job. I have to work as hard as I always have when something needed to get done."

Richard began to live most of his life inside his head. He began to distrust his people. They never seemed to work hard enough or to care about his business as much as he did. If he wanted to get something done, he usually had to do it himself.

Richard's secretary, Sarah, quit in a huff one day, frustrated by the amount of work he was asking of her. Richard was left with a desk full of papers and a telephone that wouldn't stop ringing.

Clueless about the work Sarah had done, Richard, was overwhelmed by having to pick up the pieces of a job he didn't understand. His world turned upside down. He felt like a stranger in his own business.

Why had he been such a fool? Why hadn't he taken the time to learn what Sarah did in the office? Why had he waited until now?

Ever the trooper, Richard plowed into Sarah's job with everything he could muster. What he found shocked him. Sarah's work space was a disaster area! Her desk drawers were a jumble of papers, pens, pencils, erasers, rubber bands, envelopes, business cards, and candy.

"What was she thinking?" Richard raged.

When he got home that night, even later than usual, he got into a shouting match with Anne. He settled it by storming out of the house to get a drink. Didn't anybody understand him? Didn't anybody care what he was going through?

Richard returned home only when he was sure Anne was asleep, and then slept on the couch. He left early in the morning before anyone was awake. He was in no mood for questions or arguments.

When Richard got to the office, something was very wrong. Robert, the bookkeeper, was AWOL.

Fear flashed through Richard's body like an electrical current. People were disappearing—first Sarah, and now Robert. But Robert handled the money!

He tried to reach Robert at home. No answer.

Rummaging through Robert's desk, Richard discovered that the checkbook was missing! And all the company's financial records!

Richard felt panic welling up inside of him like lava. He took several deep breaths and tried to slow his raging thoughts. Maybe Robert took the records home last night. Maybe he was putting in some overtime. He began to pace the floor, good thoughts battling it out in his mind with bad ones.

Two hours passed. Richard was beside himself. The phone kept ringing. Four jobs were in the works, and unless he visited the sites, his people wouldn't know what to do next. Still, he paced, nearly paralyzed with terror and indecision. What to do, what to do?

Finally, he called the bank. The words would forever ring in his ears: "Your balance is zero." Nightmare was now reality.

Richard was plunged to the depths of despair. "I've blown everything," he cried. "What will I say to Anne and the kids? What can I do now?"

What lessons can we draw from Richard and Anne's story? As I've already emphatically said, every business is a family business. Every business profoundly touches every family member, even those not working in the business. Every business either gives to the family or takes from the family, just as individual family members do.

If the business takes, the family is always the first to pay the price.

In the case of Richard, in order for him to free himself from the prison he created, he first had to admit his vulnerability. He had to confess to his wife and to the rest of his family that he really didn't know enough about his own business and how to grow it.

Richard, like so many other Contractors, had tried to do it all and keep it all to himself. Had he succeeded, had the business supported his family in the style he imagined, he would have burst with pride. Instead, Richard had unwittingly isolated himself, thereby achieving the exact opposite of what he sought.

He destroyed his life—and his family's life along with it.

Repeat after me: *Every business is a family business.*

Are you like Richard? I believe that all Contractors

share a common soul with him. And that all of you must learn—some the hard way—that a business is only a business. It is not your life. But it is also true that your business can have a profoundly negative impact on your life unless you learn how to do it differently than most Contractors do it. Differently than Richard did it. Perhaps differently than you do it.

The tragedy in Richard's case is that his business could have served his and his family's life. But to have made that happen, Richard would have had to learn how to master his business in a way that was completely foreign to him.

What happened instead is that Richard's business consumed him. Lacking a true understanding of the essential strategic thinking that would have allowed him to create something unique, Richard and his family were doomed before he even opened his doors.

What follows are the secrets Richard should have—could have—known if he had only been interested.

Let's start with the subject of *money*.

# CHAPTER TWO

# On the Subject of Money

Money, money, money, it's driving me crazy.

*THE POWER POINT*

Had Richard and Anne first considered the subject of money the way we are about to here, their lives would have been radically different.

Of course, money is on the tip of every Contractor's tongue, on the edge (or at the center) of every Contractor's thoughts, intruding upon every part of a Contractor's life.

With money consuming so much energy in most Contractors' businesses, why do so few of them handle it well? Why was Richard willing—as so many Contractors are—to entrust financial affairs to a relative stranger? Why is money always scarce for most Contractors? Why is there less money than we thought there would be? And yet the demand for money is *always* high.

What is it about money that is so elusive, so complicated, so difficult to control? Why is it that every Contractor I've ever met hates to deal with the subject of money? Why are they almost always too late in facing money problems? And why are they constantly obsessed with the desire for more of it?

*Money:* You can't live with it, and you can't live without it. But you better understand it. Because until you do, it will eat your business for lunch!

## THE FOUR FACTORS OF MONEY

In the context of owning, operating, developing, and exiting from a Contractor's business, four discrete, yet highly integrated, factors govern the subject of money in every business:

1. **Income**

2. **Profit**

3. **Flow**

4. **Equity**

Failure to discern how the four factors of money play themselves out in your business is a surefire recipe for disaster.

**Important Note:** Do not talk to your accountants or bookkeepers about what follows; it will only confuse them and you. This information comes from the real-life experiences of thousands of small business owners, Contractors included, most of whom were hopelessly confused about money when I first met them. Once they understood and accepted the following principles, however, they soon developed a clarity about money that could only be called enlightened.

## The First Factor of Money: Income

Income is the money that Contractors are paid by their business for doing their job *in* the business. It's what they get paid for going to work every day.

Clearly, if Contractors didn't do their job, others would have to, and they would be paid the money the business currently pays the Contractors.

Hear me. Income has nothing to do with *ownership.* Income is solely the province of *employee-ship.* That's why to the Contractor-as-*Employee,* Income is the most important form money can take. To the Contractor-as-*Owner,* in contrast, Income is the least important form money can take.

Most important; least important. Do you see the conflict taking shape? The conflict between the Contractor-as-Employee and the Contractor-as-Owner?

We'll deal with this conflict later. For now, just know that this conflict between the Contractor-as-Employee and the Contractor-as-Owner is potentially the most paralyzing conflict in a Contractor's life.

Resolving this conflict will set you free!

## The Second Factor of Money: Profit

Profit is what's left over after a business has done its job effectively and efficiently. If there is no Profit, there is something wrong with the business.

However, just because the business shows a Profit does not mean it is necessarily doing all the right things in the right way. Rather, it just means that something was done right during or preceding the period in which the Profit was earned.

The important issue here is whether the Profit was intentional or accidental. If it happened by accident (which most Profit does), don't take credit for it. You'll live to regret your impertinence.

If it happened intentionally, take all the credit you want. You've earned it. And Profit created intentionally, rather than by accident, is replicable—again and again.

As you'll soon see, the value of money is a function of your business's ability to produce it in predictable amounts at an above average Return on Investment.

Profit can be understood only in the context of the business's purpose, as opposed to the owner's purpose. Profit, then, fuels the forward motion of the business that produces it. This is accomplished in four ways:

**1. Profit is *investment capital* that feeds and supports growth.**

**2. Profit is *bonus capital* that rewards people for exceptional work.**

**3. Profit is *operating capital* that shores up money shortfalls.**

**4. Profit is *return-on-investment capital* that rewards the owner for taking risks.**

Without Profit, a business cannot subsist, much less grow. It is the fuel of progress.

If a business misuses or abuses Profit, however, the penalty is much like having no Profit at all. Imagine the plight of a Contractor who has way too much return-on-investment capital and not enough investment capital, bonus capital, and operating capital.

Can you see the imbalance this creates?

## The Third Factor of Money: Flow

Flow is what money *does* in a business, rather than what money *is*. In a small Contractor's business, money tends to move erratically through it, like a pinball. Sometimes it's there, and sometimes it's not.

Flow can be even more critical to a business's survival than Profit, because a business can produce a Profit and still be short of money. It's called Profit on Paper, rather than in fact.

And if you're a small Contractor and the money isn't there when it's needed, you're in danger, no matter how much Profit you've made. Of course, you can borrow money. But money acquired in dire circumstances is almost always the most expensive.

Knowing where the money is and where it will be when you need it is a critically important task of both the Contractor-as-Employee and the Contractor-as-Owner.

### RULES OF FLOW

You will not learn a more important lesson than the huge impact that Flow can have on the health and survival of your business. The following rules will help you understand why this subject is so critical:

**1.** The First Rule of Flow states that your Income Statement is static, while the Flow is dynamic. Your Income Statement is a snapshot, while the Flow is a moving picture. So, while your Income Statement is an excellent tool for *analyzing* your business after the fact, it's a poor tool for *managing* your business in the heat of the moment.

Your Income Statement tells you (1) how much money you're spending and where; and (2) how much money you're receiving and from where.

Flow gives you the same information as the Income Statement, plus it tells you *when* you're spending and receiving money—that is, Flow is an Income Statement moving through time. And that is the key to understanding Flow. It is about management in real time. How much is coming in? How much is going out? You'd like to know this daily, or even by the hour if possible. Never by the week or month.

You must be able to forecast the Flow. You must have a Flow Plan that helps you gain a clear vision of what's out there next month and the month after that.

You must also know what your needs will be in the future. Ultimately, however, when it comes to Flow, the action is always in the moment.

It's about now!

Managing Flow calls for incredible attention to detail. But when Flow is managed, your life takes on an incredible sheen. You're in charge! You are swimming with the current, not upstream.

2. The Second Rule of Flow states that money seldom moves as you expect it to. But you do have the power to change that provided you understand the two primary sources of money as it comes in and goes out of your practice.

The truth of it is, the more control you have over the *source* of money, the more control you have over its Flow. The sources of money are both inside and outside of your business.

Money comes from *outside* your business in the form of receivables, sales, investments, and loans.

Money comes from *inside* your business in the form of payables, taxes, capital investments, and payroll. These are the costs associated with marketing, sales, operations, cost of goods, and so forth.

Many Contractors don't see money going out of their business as a source of money, but just the opposite is true.

When considering how to spend money in your business, you can save—and therefore make—money in three ways:

1. **Do it more effectively.**

2. **Do it more efficiently.**

3. **Stop doing it altogether.**

By identifying the money sources inside and outside of your business, and then applying these methods, you are better able to control the Flow in your business.

But what are these sources? They include how you:

- Plan a project

- Buy materials

- Compensate your people

- Plan people's time

- Estimate a job

- Sell a job

- Manage a job

- Collect receivables

And countless more. In fact, every task in your business can be done more efficiently and effectively. In the process, you will create more Income, produce more Profit, and balance the Flow.

### The Fourth Factor of Money: Equity

Sadly, few Contractors will ever fully appreciate the value of Equity in their business. Yet, Equity is the second most valuable asset any Contractor will ever possess. (The first most valuable asset is, of course, your life, we'll talk more about that later.)

So let me define the word *Equity* for you:

> **EQUITY** is the financial value placed on your business by a prospective buyer of your business.

Thus, your *business* is your most important product, not your services. Because your business has the power to set you free. That's right! Once you sell your business—providing you get what you want for it—you're free!

Of course, to enhance your Equity, to increase your business's value, you have to build it right. You have to build a business that works. A business that can produce Income, Profit, Flow, and Equity better than any other Contractor's business can.

To accomplish that, your business must be designed so that it can do what it does systematically, predictably, every single time.

## The McDonald's Story

Ray Kroc called his first McDonald's restaurant "a little money machine." That's why thousands of franchisees bought it. And the reason it worked? Ray Kroc demanded consistency. So that a hamburger in Philadelphia would be an advertisement for one in Peoria. In fact, no matter where you bought a McDonald's hamburger in the 1950s, the meat patty was guaranteed to weigh exactly 1.6 ounces, with a diameter of $3^5/8$ inches. It was in the McDonald's handbook.

Did Ray Kroc succeed? You know he did! And so can you, once you understand his methods. Consider just one part of Ray Kroc's story.

In 1954, Ray Kroc made his living selling the five-spindle Multimixer milkshake machine. He heard about a hamburger stand in San Bernardino, California, which had eight of his machines in operation, meaning it could make 40 shakes simultaneously. That he had to see.

Kroc flew from Chicago to Los Angeles, then drove 60 miles to San Bernardino. As he sat in his car outside Mac and Dick McDonald's restaurant, he watched as lunch customers lined up for bags of hamburgers.

In a revealing moment, Kroc approached a strawberry blonde in a yellow convertible. As he later explained, "It was not her sex appeal but the obvious relish with which she devoured the hamburger that made my pulse begin to hammer with excitement."

*Passion.*

In fact, it was the french fry that truly captured his heart. Before the 1950s, it was almost impossible to buy fries of consistent quality. Ray Kroc changed all that. "The french fry," he once wrote, "would become almost sacrosanct for me, its preparation a ritual to be followed religiously."

*Passion and preparation.*

The potatoes had to be just so—top-quality Idaho russets, 8 ounces apiece, deep-fried to a golden brown, and salted with a shaker that, as Kroc put it, kept going "like a Salvation Army girl's tambourine."

As Kroc soon learned, potatoes too high in water content—and even top-quality Idaho russets varied greatly in water content—will come out soggy when fried. And so Kroc sent out teams of workers, armed with hydrometers, to make sure all his suppliers were producing potatoes in the optimal solids range of 20 to 23 percent.

Preparation and passion. Passion and preparation. Look those words up in the dictionary, and you'll see Ray Kroc's picture. Can you see your picture there?

Do you understand what Ray Kroc did? Do you see why he was able to sell thousands of franchises? He knew the true value of Equity, and, unlike Richard, Kroc went to work *on* his business, rather than *in* his business.

So what's your business version of that statement? What does your Contractor's business need to do to become a little money machine? What is the passion that will drive you to build a business that works, a turnkey system like Ray Kroc's?

## Equity and the Turnkey System

What's a turnkey system? And why is it so valuable to you? To better understand it, let's look at another example of a turnkey system that worked to perfection: the recordings of Frank Sinatra.

Frank Sinatra's records were to him exactly what McDonald's restaurants were to Ray Kroc. They were part of a turnkey system that allowed Sinatra to sing to millions of people without having to be there himself.

In this way, Sinatra's recordings were a dependable turnkey system that worked predictably, systematically, automatically, and effortlessly to produce the same results every single time.

No matter where Frank Sinatra was, his records just kept on producing Income, Profit, Flow, and Equity, over and over. Sinatra needed only to produce the proto-type recording, and the system did the rest.

Similarly, Kroc's McDonald's is also the prototypical turnkey solution, addressing everything McDonald's needs to do in a basic, systematic way so that anyone properly trained can successfully reproduce the same results.

And that's where you'll realize your Equity Opportunity: in the way your business does business; in the way your business systematically does what you intend it to do; in your turnkey system that works, even in the hands of ordinary people, to produce extraordinary results.

Remember:

- If you want to build vast Equity in your business, then go to work *on* your business. Build a business that works every single time.

- Go to work *on* your business to build a totally integrated turnkey system that delivers every single time.

- Go to work *on* your business to package it and make it stand out from the Contractors' businesses you see everywhere else.

Here is the most important idea you will ever hear about your business and what it can potentially provide for you:

**The value of your Equity is directly proportional to how well your business works. And how well your business works is directly proportional to the effectiveness of the systems you have put into place, upon which the operation of your business depends.**

Money . . . Happiness . . . Life . . . they all come down to how well your business works.

Whether money takes the form of Income, Profit, Flow, or Equity, the amount of it—and how much of it stays—will always depend on how well your business works.

Your business holds the secret to more money. Are you ready to learn how to find it?

Earlier in this chapter, I alerted you to the inevitable conflict between the Contractor-as-Employee and the Contractor-as-Owner. Between the part of you working *in* the business and the part of you working *on* the business. Between the part of you working for Income and the part of you working for Equity.

Here's how to resolve this conflict:

1. Acknowledge to yourself when you are filling employee shoes and when you are filling owner shoes.

2. As employee, determine the most effective way to do the job you're doing, *and then document that job.*

3. Once you've documented the job, create a strategy for replacing yourself with someone else, who, after learning how to faithfully use the system you've provided, will then teach it to yet another person.

4. Manage the newly delegated system using your new employees. Improve the system by quantifying its effectiveness over time.

5. Repeat the above process throughout your business wherever you are acting as employee rather than owner.

6. Leave behind dedicated people using your effective systems, each time moving you out of employee-ship work and freeing you to do ownership work.

Master these methods, understand the difference between the Four Factors of Money, develop an interest in how money works in your business, and then watch it flow in!

# CHAPTER THREE

# The Ethical Home™:
# One Contractor's Story

*What is impossible to do, but if it could be done, would*
*fundamentally change your business?*

*PARADIGMS*
JOEL ARTHUR BARKER

The following story was written by a real home builder, a man motivated by The E-Myth to create a completely new kind of home-building business. He wanted a business that would serve customers, clients, and employees in an uncommonly human way. I applaud this home builder not only for the story he allowed me to share with you, but also for his heartfelt commitment to it. It's "The Story of The Ethical Home."

## THE STORY OF THE ETHICAL HOME

*Mike and Mary had saved for a long time to*
*build their dream house, the largest investment*

of their lives. They often talked late into the night about what the house would look like. They agreed on the number of bedrooms and bathrooms, the location of the fireplace, and how much closet space they would need. Decisions were reached about patios, porches, kitchen cabinets, and appliances. They knew exactly what they wanted in their dream house, and now, finally, they were ready.

Mike and Mary took care in selecting a builder, because they knew full well it was the most important decision in their quest for a dream house. They contacted the Better Business Bureau, the local Contractors' association, talked to friends, and drove around on weekends looking at dozens of houses.

They finally chose Expert Construction Company, owned and operated by Jim Adams, who had been building in the area for two decades. Mike liked the way Jim constructed houses. He seemed to go out of his way to make sure his houses were solidly built. And Jim seemed in total control of every facet of his projects. Mike noticed that when Jim wasn't present, not much happened at his job sites, although this was seldom a problem because Jim always seemed to be around. For her part, Mary liked his sense of color and choice of materials. No doubt about it, Jim and Expert Construction were a good fit.

Mike and Mary worked with a house-planning firm recommended by Jim, and produced just the design they had envisioned. Everything was moving as well as they had hoped.

Soon after construction began, however, questions arose. Mary wanted to know why the concrete trucks had made such deep ruts in the yard and why they had to spill patches of cement everywhere. "We'll never get that removed completely," she thought. When she asked Jim, he assured her it would be okay. And the holes in the side of the slab face? "Not to worry," Jim cooed. "It's a normal occurrence."

Now Mike was concerned, too. The plumbing pipes located in the slab did not seem to have enough fall. And the plumbing pipes protruding from the slab—they were now several inches from where the plans said they should be. That meant the affected wall locations had to be changed, which would in turn change the room sizes. Jim said this was not uncommon and not to worry about it.

More and more problems came to their attention. When the insulation was installed, some batts were not secured, and Mike was afraid that eventually they would fall behind the wall. Another time the Sheetrock hangers drove a nail through a water line. The Sheetrock had to be torn out, the water line repaired, and the wall patched. "You won't even notice the patch," Jim assured them. Then they discovered spots where the paint didn't completely cover the walls and ceiling. Jim called them "holidays" and once again reassured Mike and Mary that he would take care of it.

When Mary saw the newly painted master bedroom, she was angry. No way had she picked out that color! She was shocked to learn that it was the right color. Jim explained that when a small color sample is multiplied a thousand times to cover a room, it can look quite different.

By now, Mike and Mary were more than a little disillusioned with Expert Construction Company. Jim, seemingly weary of the constant strain, was less friendly and cooperative. When Mike and Mary learned that the job was going to run over cost by $5,000, Jim insisted it was due "to the change orders they had requested." Their blood pressure rose even more when they learned the completion date would be delayed six weeks, partly due to those change orders and partly due to Sub-Contractor problems and bad weather.

Finally, the dream house was finished, and Mike and Mary moved in. Not surprisingly, many little aggravating, "nitpicking" (Jim's word) problems surfaced: a piece of molding was missing in the laundry room, paint was spotty, a cabinet door in the kitchen wouldn't close, the upstairs shower made a noise when it was turned on, and there never seemed to be enough hot water.

Jim grew reluctant to return to take care of the

*problems. Tension and frustration defined the relationship between the couple and the Contractor. Mike and Mary concluded they would just have to learn to live with the little frustrations of home ownership.*

The home builder who wrote this story added the following summary. As you read it, imagine you're a prospective buyer meeting with this person. Imagine he has just shared "The Story of The Ethical Home" with you:

*Having been in the building business for over 30 years, I've seen the Mike, Mary, and Jim story played out all too often. It boils down to Contractors showing little regard for the needs of the home buyer and the home buyer showing little interest in how to make a wise home-buying decision.*

*Such a home builder's philosophy can be reduced to one command: "Get the job done, full speed ahead, no matter what the problems! We'll deal with them when they happen!"*

*In most cases, quality ends up coming in a distant second to the bottom line. Customer concerns are given low priority. This misguided emphasis inevitably produces poor workmanship, poor cost control, poor quality control, and miserable Customer relations. I've seen it happen over and over.*

*But it didn't have to happen. The desire to make sure it didn't happen was the genesis of my company, the Classic Development Corporation, and my concept of The Ethical Home.*

Let's return to our story to see what could have been done to produce a more satisfying result for everyone concerned:

*Of course, Mike and Mary could have done things to make sure Jim fulfilled his promises, but we'll explore them later. For the moment, let's concentrate on Jim and Expert Construction Company.*

*You should realize that Jim is an honest person who wants to do the best job he can for his Customers. So what went wrong? Why couldn't he fulfill the expectations of Mike and Mary?*

*To begin with, Jim is a technician and not a businessman. He knows building, and he knows what needs to be done to produce a home. But— and this is often the case with Contractors—Jim has only his technical knowledge, goodwill, and good intentions to see him through the job.*

*Unfortunately, as we've already seen, good intentions are simply not enough. Jim is not in control of his business. How do we know this? Nothing seems to get done or run smoothly unless he's on the job. Jim doesn't have the systems in place to ensure that his Sub-Contractors will do the job and do it right even in his absence. He hasn't made it clear to his Sub-Contractors what standards they must adhere to when doing their work. He has become, in effect, their employee. No system is in place for him to check their work with only a minimum investment of Jim's time. Because he has no system, they have no system. And because his Sub-Contractors have no system, Mike and Mary are left to depend on good luck.*

*And when things began to go wrong, as they inevitably do on any construction project, Jim had no plan for handling the changes requested by Mike and Mary. He should have addressed every aspect of the home project before construction began and incorporated the changes into the original plan.*

*Instead, when unforeseen circumstances occurred, Jim had no system in place to deal with them. His failure to communicate meant that Mike and Mary were unprepared for extra costs and time delays.*

*Jim should have had a thorough preconstruction conference with Mike and Mary and gone over every stage of construction. This would have clarified what the couple could expect to see as the job progressed. Instead, no time was devoted to addressing their expectations or to adjusting those expectations when reality intruded. Later, Jim didn't carefully guide Mike and Mary through the finished home and demonstrate how everything worked.*

*Remember, there are over 30,000 different components that go into a typical home today. It should never be the owners' responsibility to understand them all.*

*Finally, Jim neither spelled out his warranted responsibilities clearly nor created an orderly system to handle the all-important final punch-out.*

*Like most Contractors, Jim was out of control. He was out of control personally, which was reflected in his inability to control his business. To the contrary, his business was actually controlling him.*

*Without a clear definition of what his business really was, without strategic goals, Jim was just building a house. What Jim needed was to understand clearly his purpose in building homes in the first place. Why was he in the construction business instead of, say, the restaurant business? Only when Jim found such clarity would he be able to focus on how to achieve the results needed to satisfy his purpose.*

"It was with this in mind," our home builder said, "that I formed the Classic Development Corporation. Solving the problems that Jim, Mike, and Mary encountered ultimately led to the creation of The Ethical Home, which begins with my company's purpose statement." The purpose statement is:

*Classic Development Corporation's purpose is to serve the home-buying public, to make sure the home-buying and ownership experience is joyful, easy, and rewarding. This means:*

- *Dealing with you, the Customer, ethically and honestly*

- *Producing a quality product that you can depend on to look and function exactly as promised*

- *Eliminating all sources of frustration*

- *Using the best, most efficient management controls available*

- *Realizing a fair and reasonable profit at a reasonable cost to you*

*To get the results that both the Customer and my company, Classic Development Corporation, wanted and needed, I realized I first had to address the issue of ethics. I needed to commit to treating everyone in my business sphere with the respect and dignity they deserve.*

*I further committed to dealing only with Sub-Contractors and suppliers who could embrace the same high standards as my company. I made sure that the Customers' needs were clearly identified from the outset, and ultimately met. I prepared suppliers, Sub-Contractors, and home buyers for what I expected throughout the home construction process.*

*So I set written standards of work quality and conduct for everyone who works with and for Classic Development. I demanded that people do what they say they're going to do, such as com-*

*pleting the job in a timely manner, controlling costs, and offering the highest-quality workmanship through efficient management systems.*

*In summary, it was and is my belief that taking an ethical approach to building a home will inevitably produce trust, confidence, and satisfaction. It will create something extraordinary—The Ethical Home—which will leave the homeowner frustration-free throughout the entire process.*

*One more thing. I told you before that Mike and Mary could have done something to avoid the frustration they encountered. The answer is simple: From the outset, they should have looked for a company dedicated to the principles I've just described to you. They should have looked for* real *professionals like those at Classic Development Corporation.*

*They should have looked for a company with a Systems Approach to Complete Customer Satisfaction!*

The preceding story is true. It's an account created by a Contractor who understands the power of a story to touch the heart of his Customers, employees, suppliers, lenders, and Sub-Contractors. This Contractor knows what the true purpose of his business really is.

What's the purpose—the story—of *your* business?

Although you may think you are in the electrical contracting, the plumbing contracting, or the remodeling business, like every other truly entrepreneurial business founder, in reality you are in the business of business. No matter what that business is, you must understand it and it must follow certain fundamental rules if it is to be a faithful expression of your vision.

To succeed, your business must have at its foundation a story that drives and directs it day by day.

Again, what is the story of your business? And what does it *need* to be to distinguish it from other Contractors' businesses? What's *your* version of The Ethical Home? And once you envision it, how can you make it come true?

# CHAPTER FOUR

# On the Subject of Planning

*Freedom does not come automatically; it is achieved. And it is not gained at a single bound; it must be achieved each day.*

*MAN'S SEARCH FOR HIMSELF*
ROLLO MAY

A nother obvious oversight revealed in Richard and Anne's story was the absence of true planning.

It goes without saying, but we're here to say it anyway . . . *every Contractor must have a plan.* You can't even begin a job without one. But, like Richard, most Contractors fail to understand that The Job Plan is only one of *three* critical Plans they need to create and then implement to avoid failure. These three Plans are The Business Plan, The Job Plan, and The Completion Plan.

Together these three Plans form a triangle, with The Business Plan at the base, The Job Plan in the center, and The Completion Plan at the apex.

# THE PLANNING TRIANGLE

The Business Plan determines *who* you are (the Business), The Job Plan determines *what* you do (the Job), and The Completion Plan determines *how* you do it (the Fulfillment Process).

By looking at the Planning Triangle, we see that the Three Critical Plans are interconnected. The connection between them is established by asking the following questions:

**1. Who are we?**

**2. What do we do?**

**3. How do we do it?**

*Who are we?* is purely a Strategic Question.

*What do we do?* is both a Strategic and a Tactical Question.

*How do we do it?* is both a Strategic and a Tactical Question.

First ask: "What do we do and how do we do it . . . *strategically?*"

And then: "What do we do and how do we do it . . . *tactically?*"

The Strategic Questions shape the vision and the destiny of your business. The Tactical Questions turn that vision into reality. Thus, Strategic Questions provide the foundation for Tactical Questions, just as the base of the Triangle provides the foundation for the middle and apex.

Let's take a brief look at each one of the three Plans and what they will do for you in the development of your business.

## THE BUSINESS PLAN

Your Business Plan will determine what you choose to do in your business and the way you choose to do it. Without a Business Plan, your business can do little more than survive. And without a Business Plan, even basic survival will require luck.

Without a Business Plan, you're treading water in a deep pool, with no shore in sight. You're working against the natural flow.

But I'm not talking about the traditional Business Plan that is taught in business schools. No, the Business Plan I'm talking about reads like a story, the most important story you've ever told.

Your Business Plan should clearly describe:

- The business you are creating

- The Purpose it will serve

- The Vision it will pursue

- The process through which you will turn that Vision into a Reality

- The way the money works as your business grows to realize your Vision

This process of moving your business from where it is today to where you see it in the future is defined by your Business Benchmarks—the goals you want your business to achieve during its lifetime.

Your Business Benchmarks will include Financial Benchmarks; Emotional Benchmarks (the visual, emotional, functional, and financial impact your business will have on everyone who comes into contact with it); Performance Benchmarks; Customer Benchmarks (Who are they? Why will they buy from you? What will your business give them that no one else's will?); Employee Benchmarks (How do you grow people? How do you find people who want to grow? How do you create a school in your business that will teach your people skills they can't learn anywhere else?); and many, many more.

Your Business Benchmarks will reflect the position your business will hold in the minds and hearts of your Customers and your employees, your investors and your suppliers, and how you intend to make that position a reality through the systems you develop.

Your Business Benchmarks will describe how your management team will take shape, and what systems you will need to develop so that your managers, just like McDonald's managers, will be able to produce the results for which they will be held accountable.

And finally, your Business Plan will be built with *Business* language, not *Contractors'* language. It will detail the Market and the Strategy through which you intend to become a leader in that Market. It will rely on Demographics and Psychographics—that is, who buys and why they buy. It will include Return on Investment and Return on Equity. It will focus on the things in which your lenders and shareholders are interested. Not the things in which your technicians are interested.

A Contractor lacking both a business vision and the words to articulate it is simply a guy who goes to work every day. A guy like every other guy who's just doing it, doing it, doing it, Busy, busy, busy. Maybe making money, maybe not. Maybe getting something out of life, maybe not. Taking chances, without ever really taking control.

Therefore, the primary purposes of the Three Critical Plans are to clarify precisely what needs to be done to get what the Contractor wants from his or her business and life, and to define the specific steps by which this will happen.

First this must happen. Then that must happen. One. Two. Three. One step at a time.

Let's take a look at The Job Plan.

The Job Plan includes everything a Contractor needs to know, have, and do in order to complete a job on time, every time, exactly as promised.

Every job should prompt you to ask three questions:

1. **What do I need to know?**

2. **What do I need to have?**

3. **What do I need to do?**

## What Do I Need to *Know*?

What information do I need to complete this job on time, every time, exactly as promised? In order to know what you need to know, you must understand the expectations of others: the Customer, the architect, the general Contractor, Sub-Contractors, and your employees. Are you clear on those expectations? Don't assume you know. Instead, create a Need-to-Know Checklist to make sure you ask all the necessary questions.

## What Do I Need to *Have*?

This question raises the issue of resources—namely Money, Equipment, People, and Time. If you don't have the right equipment, how are you going to complete the job? If you don't have enough money to finance the job, how can you expect to complete it without creating cash flow problems? If you don't have sufficient people or people with sufficient skills, what happens then? And if you don't have enough time to manage the job to completion, what happens when you have to be in two places at once?

Again, don't assume that you can get what you need when you need it; most often you can't. And even if you can get what you need at the last minute, you will pay a dear price for it.

## What Do I Need to *Do*?

The focus here is on actions to be started and finished. What do I need to do to get this job done on time, every time, exactly as promised? Answering this question demands a series of Action Plans, including:

- The Objective to be achieved

- The Standards by which you know the Objective will be achieved

- The Benchmarks you need to reach in order for the Objective to be achieved

- The function/person accountable for the completion of the Benchmarks

- The Budget for the completion of each Benchmark

- The Time by which each Benchmark must be completed

The Action Plans you create become the foundation for the Completion Plans you need to create in order to assure that everything is not only realistic but can be managed.

## THE COMPLETION PLAN

If The Job Plan tells you in advance everything you need to know about the job, The Completion Plan tells you everything you need to know about every Benchmark in The Job Plan. The goal: to fulfill your commitment to complete the job in the way and in the time you promised.

The Completion Plan provides for information about the details of doing Tactical Work. It is a guide to tell the people responsible for doing that work exactly how to do it. Every Completion Plan becomes a part of the knowledge base of your business. No Completion Plan goes to waste. Every Completion Plan becomes a textbook for new employees, Sub-Contractors, architects, and Customers to help explain how your company does business in a way that distinguishes it from the businesses of all other Contractors.

For example, at McDonald's The Completion Plan for making a Big Mac is explicitly described in *McDonald's Operations Manual,* as is every single Completion Plan needed to run a McDonald's business.

The Completion Plan for a Contractor might

include how you lay a foundation, in contrast to how everyone else does it. How you answer that question will determine exactly how you distinguish your business from every other Contractor's business.

## BENEFITS OF THE PLANNING TRIANGLE?

Can you appreciate the impact that implementing a Planning Triangle would have on your business? On your life?

By doing exactly what I've described above, you will find out:

- What your business is going to look, act, and feel like when it's finally done

- When it's going to happen

- How much money you will make

And much, much more.

Imagine how valuable it would be to be able to monitor your progress, step-by-step; to know for certain whether you're on the right track.

Well, that's what Planning is all about. It's about creating a standard—a yardstick—against which you will be able to measure your performance.

Failing to create such a standard is like throwing a straw into a hurricane. Who knows where that straw will land?

But a Contractor with a Plan! Who wouldn't want to do business with such a person?

(To find out exactly what your three critical Plans will look like when they're done, go to E-Myth.com and click on The E-Myth Contractor.)

# CHAPTER FIVE

# On the Subject of Management

> There is always an unknown quality in the
> creative process, as there is in fishing, but when you are aware
> of the final result you want to create, you are able to focus
> the process, rather than make the process a random one.
>
> *THE PATH OF LEAST RESISTANCE*
> ROBERT FRITZ

Every Contractor, including Richard and the builder of The Ethical Home, eventually faces the issue of management. Yet most of us face it so badly!

Why do so many Contractors suffer from a kind of paralysis when it comes to dealing with management? Why are so few Contractors able to get the job done the way they promised, and on time? Why are their managers (if they have any) so inept?

It's because we are trying to manage people contrary to how we need to do it.

Let me explain. We often hear that a good manager must be a "people person." Someone who loves to nourish, figure out, support, care for, teach, baby, monitor, mentor, direct, track, motivate, and, if all else fails, threaten or beat up his or her people.

Don't believe it. Because management has far less to do with people than you've been led to believe.

In fact, despite what every management book written by management gurus (who seldom have managed anything) says, no one yet—besides a few tyrants—has ever learned how to manage people.

And the reason is simple: *People are unmanageable!*

Yes, it's true. People are unmanageable. What's more, they are inconsistent, unpredictable, unchangeable, unrepentant, irrepressible, and generally impossible.

Doesn't knowing this make you feel better? Now you understand why you've had all those problems! Do you feel the relief, the heavy stone lifted from your chest?

The time has come to fully understand what management is really all about. Rather than managing *people,* management is really all about managing a *process,* a step-by-step way of doing things, which, combined with other processes, becomes a system. For example:

- The process for completing an estimate

- The process for answering the telephone

- The process for installing a switch plate

- The process for erecting a wall

Thus, a process is the step-by-step way of doing something over time. Considered as a whole, these processes are a system:

**The Estimating System**

**The Telephone Answering System**

**The Switch-Plate Installation System**

**The Wall-Erecting System**

Instead of managing people, then, the truly effective manager is one who has been taught a system for managing a process through which people get things done.

More precisely, managers and their people, *together,* manage the processes—the systems—that compose your business. Management, we now see, is less about *who* gets things done in your business than about *how* things get done.

Great managers are fascinated with how things get done. They are masters at figuring out how to get things done effectively and efficiently.

Great managers constantly ask key questions:

- What is the result we intend to produce?

- Are we producing that result every single time?

- If we're not producing that result every single time, why not?

- If we are producing that result every single time, how could we produce even better results?

And so forth.

In short, great managers are those who use a great management system. A system that fairly shouts out, "This is *how* we manage here!" Not, "This is *who* manages here!"

In a truly effective company, how you manage is always more important than who manages. Because how you manage is transferable, whereas who manages isn't. How you manage can be taught, whereas who manages can't be taught.

In short, when a company is dependent upon who manages—Murray, Mary, or Moe—that company is in serious jeopardy. Because when Murray, Mary, or Moe leaves, the company has to start over again. What an enormous waste of resources!

Even worse, when a company is dependent upon who manages, you can bet all the managers in that company are doing their own thing. What could be more unproductive than having 10 managers, each managing in a unique way? How in the world could you possibly manage those managers?

The answer is clear: You couldn't. Because it takes you right back to trying to manage people again.

And, as we now know, that's impossible!

What, then, is a Management System?

The E-Myth says that a Management System is the

method by which every manager innovates, quantifies, orchestrates, and then monitors the systems through which your business produces the results you expect.

According to The E-Myth, a manager's job is simple:

**A manager's job is to invent the systems through which the owner's vision is consistently and faithfully manifested at the operating level of the business.**

Which brings us right back to the purpose of your business, to the need for a Vision, and to "The Story of The Ethical Home."

Are you beginning to see what I'm trying to share with you? That your business is one single thing. And that all the subjects we're discussing here—money, planning, your company's story, management, and so on—are all about doing one thing well. The one thing your business is intended to do.

And it's the manager's role to make certain it works.

# CHAPTER SIX

# On the Subject of People

First he told me what he was going to do,
then he did something else. But he actually
thought he had done what he told me
he was going to do! Who can explain it?

ANONYMOUS CONTRACTOR

Every Contractor I've ever met has complained about people.

About employees: "They come in late, they go home early, they're smoking dope at lunch!"

About Sub-Contractors: "Who knows what they do with their time—but they certainly know how to charge for it!"

About suppliers: "They're living in a nonparallel universe!"

About their customers: "Even if they had a mind, they wouldn't be able to make it up!"

People, people, people. Every Contractor's nemesis. And at the heart of it all are the people who work for you.

Have you ever heard yourself say any of the following? "By the time I tell them how to do it, I could have done it 20 times myself!" "How come nobody listens to what I say?" "Why is it nobody ever does what I ask them to do?"

Working with people brings great joy—and monumental frustration. And so it is with Contractors and their people. But why? And what can we do about it? Let's look at the typical Contractor. Who this person is—and who he or she isn't. As you read the following, consider Richard, the Contractor whose story I told in the beginning.

Few Contractors are prepared to use other people to get results. Not because they can't find people, but because they are fixated on getting the results themselves! In other words . . . most Contractors are not the businesspeople they need to be, but technicians suffering from an entrepreneurial seizure!

Am I talking about you? What were you doing before you became a Contractor?

If you are a plumbing Contractor today, were you a plumber yesterday?

If you are an electrical Contractor today, were you an electrician yesterday?

If you are a general Contractor today, were you a carpenter yesterday?

Didn't you imagine owning your own business as the way out?

Didn't you think that because you knew how to do

the technical work—electrical, plumbing, carpentry—
that you were automatically prepared to create a business that does that type of work?

Didn't you figure that by creating your own business, you could dump the boss once and for all? How else to get rid of that impossible person, the one driving you crazy, the one who never let you do your own thing, the one who was the main reason you decided to take the leap into a business of your own in the first place?

Didn't you become a Contractor so that you could become your own boss?

And didn't you imagine that once you became your own boss, you would be free to do whatever you wanted to do—and to take home *all* the money?

Didn't you think you could get beyond the weekly payroll check?

Honestly, isn't that what you imagined? So you went into business for yourself and immediately dove into work? Doing it, doing it, doing it. Busy, busy, busy.

Until one day you realized (or maybe not) that you were doing all of the work? You were doing everything you knew how to do, plus a lot more you knew nothing about. Building sweat equity . . . you thought.

In fact, a technician suffering from an entrepreneurial seizure! Just hoping to make a buck in your own business. And sometimes you did earn a wage, but other times you didn't. You were the one signing the checks all right, but almost all of them went to other people.

Does this sound familiar? Is it driving you crazy?

Well, relax, because we're going to show you the right way to do it this time.

Read carefully. Be mindful of the moment. You are about to learn the secret you've been waiting for all your working life.

## THE PEOPLE LAW

It's critical to know this about the Contractor's Game: *Without people you don't own a business, you own a job.* And most often it's the worst job in the world . . . *because you're working for a lunatic!* (Nothing personal—we simply have to face the facts if we're ever going to change.)

The sad truth is that without people, you're going to be forced do it all yourself. Without other people, you're doomed to try to do too much. You end up knocking yourself out, 12 to 16 hours a day. You try to do more, but less actually gets done.

The load can double you over, leave you panting. In addition to the work you're used to doing, when you own your own business you're also going to have to do the books. And the estimating. And the selling. And the planning, the buying, the ordering, and the picking up. In a small contracting business of your own, the work is endless—as I'm sure you've found it—and until you discover how to get it done by somebody else, it will turn you into a burned-out husk.

Like painting the Golden Gate Bridge, it's endless.

Which puts it beyond the realm of human possibility. No matter how willing you are to do it.

But with people helping you, my Contractor friend, things can improve dramatically. If, that is, you truly understand how to engage people in the work you need them to do. When you learn how to do that, when you learn how to replace yourself with other people—*people trained in the system*—then your business can really begin to grow. And only when your business is free to grow will you begin to experience true freedom yourself.

But as I've said, you must grow it right. You must understand what it takes to see people as your greatest asset rather than your greatest liability. You must come to understand and appreciate The People Law:

> **The People Law says that each time you add a new person to your business using an intelligent (turnkey) system, a system that works, you expand your reach. And you can expand your reach almost infinitely!**

Put another way, people are to a Contractor what a record was to Frank Sinatra.

A Frank Sinatra record could be played (and still is) in a million places all at the same time, regardless of where Frank was or is. And every Frank Sinatra record sale produced royalties for Frank, and now his estate.

With the help of other people, Sinatra created a quality recording that faithfully replicated his unique talents, and then made sure it was marketed, distributed, and the revenue managed.

And as certainly as it worked for Frank Sinatra, your people can do that for you. All you need to do is to create a "recording"—a system—of your unique talents, your special way of doing everything that needs to be done in your business, and then replicate it, market it, distribute it, and manage the revenue.

Isn't that what successful businesspeople do? Make a "recording" of their most effective ways of doing business? And by doing so, they provide a turnkey solution to their customers' problems. A system solution that really works.

Doesn't your business offer the same potential for you that Frank Sinatra's records did for him (and now for his heirs)? The ability to produce income without having to go to work every day?

Isn't that what your people could be for you? The means by which your system for doing business could be faithfully replicated?

But first you've got to have a system. You have to create a unique way of doing business that you can teach to your people, that you can manage faithfully, that you can replicate consistently, just like McDonald's.

Because without such a system, without such a "recording," without a unique way of doing business that really works, all you're left with is people doing their own thing. And their own thing is almost always anarchy. Chaos. Confusion. Mistake after mistake after mistake.

And isn't that how the problem started in the first place? People doing whatever they perceived they needed to do, regardless of what you wanted? People left to their own devices, with no regard for the costs of their behavior? The costs to you?

In other words, people without a system.

Please hear this: The People Law is unforgiving. Without a systematic way of doing business, people are more often a liability than an asset. The People Law says that without a specific system for doing business; without a specific system for recruiting, hiring, and training your people to use that system; and without a specific system for managing and improving your systems, your business will always be a crapshoot.

Do you want to roll the dice with your business at stake? Unfortunately, that is what most Contractors are doing.

The People Law says that you can't effectively delegate your responsibilities unless you have something specific to delegate. And that something specific is a way of doing business that works!

Sinatra is gone, but his voice is everywhere, and someone is still counting his royalties. And that's because Sinatra had a system that works!

Do you?

# CHAPTER SEVEN

# On the Subject of Sub-Contractors

Pinch me. Is this really happening?

As long as we're on the subject of people and management, let's deal with the subject of Sub-Contractors.

Imagine this nightmare scenario: One lunatic hires another lunatic to do a job on time, and then flies into a rage when the second lunatic doesn't get the job done!

Do you sense the problem here? Do you see what's coming next?

Long ago, I've been told, God said, "Let there be Contractors. And so they never forget who they are in My creation, let them be damned forever to hire people exactly like themselves." God called those people *Sub*-Contractors.

It's educational to look up the definition of *Sub*-

*Contractor* to discover clues as to why Contractors always find themselves in predicaments. *Webster's Collegiate Dictionary, 10th Edition,* defines *sub* as "under, beneath, below"; "inferior to." *Sub-Contractor* is defined as "an individual or business firm contracting to perform part or all of another's contract."

In other words, you, the Contractor, hire someone "inferior" to you—the Sub-Contractor—to perform part or all of your contract. You make a conscious decision to hire someone "inferior" to you to fulfill your commitment to your Customer, for which you are ultimately and solely liable!

Why in the world do we do these things to ourselves? Where will this madness lead?

It seems to me that the blind are leading the blind, and the blind are paying others to do it! (Again, please don't take this personally. But, please, *do* listen to it!)

Talk about an approach doomed to fail at the very outset! It's time to step out of the darkness. It's time to see the world as it really is. It's time to do things that work.


### SOLVING THE SUB-CONTRACTOR PROBLEM

Let's say you're about to hire a Sub-Contractor. Someone who has specific skills: roofing, framing, electrical, whatever. If you've never worked with this person before, how do you know he or she is skilled? For that matter, what does skilled mean?

For you to make an intelligent decision about this

Sub-Contractor, you must have a working definition of the word *skilled.*

What if you are a Contractor who prides himself or herself on exact performance? How would you define skilled? You would certainly define it more exactly than Contractors who don't even believe in exactness.

Your challenge is to know *exactly* what you expect of the word *skilled,* and then to make sure that your Sub-Contractors operate with the same expectations. Failure here almost assures a breakdown in your relationship.

Of course, it all starts with choosing the right Sub-Contractors. After all, these are the people to whom you are delegating your responsibility, and for whose behavior you are completely liable. Do you really want to leave that choice up to chance? Are you that much of a gambler? I doubt it.

To make an intelligent selection, you need to clarify what you mean by skilled. I want you to write it down on a piece of paper: "By *skilled,* I mean . . ." Once you create your personal definition, it will become a standard for you and your business, for your Customers, and for your Sub-Contractors.

A standard, according to *Webster's 10th,* is something "set up and established by authority as a rule for the measure of quantity, weight, extent, value, or quality."

Thus, your goal is to establish a measure of quality control, a standard of skill, which you will apply to all your Sub-Contractors. More important, you are also setting a standard for the performance of your company!

By creating standards for your selection of Sub-Contractors—standards of skill, performance, integrity, financial stability, and experience—you have begun the powerful process of building a business that can operate exactly as you expect it to!

By carefully thinking about what you expect—*exactly*—you have already begun to improve your business.

In this enlightened state, you will see the selection of a Sub-Contractor as an opportunity for you to define what you (1) intend to provide for your Customers; (2) expect from your employees; and (3) demand for your life.

Powerful stuff, isn't it? Are you up to it? Are you ready to feel your rising power?

Don't rest on your laurels just yet. Defining standards is only the first step you need to take. The second step is to create a Sub-Contractor Development System.

A Sub-Contractor Development System is an action plan designed to tell you what you are looking for in a Sub-Contractor. It includes the exact standards, benchmarks, accountabilities, timing of fulfillment, and budget you will assign to the process of looking for Sub-Contractors, identifying them, recruiting them, interviewing them, training them, managing their work, auditing their performance, compensating them, reviewing them regularly, and terminating or rewarding them for their performance.

And all of these factors must be *documented*—that is, written—in what we at E-Myth like to call the Contractor's Bible™ if they're going to make any difference to you, your Sub-Contractors, your managers, or your bank account!

And then you've got to persist with that system come hell or high water.

Just as Ray Kroc did. Just as Walt Disney did. Just as Sam Walton did.

# CHAPTER EIGHT

# On the Subject of Estimating

Discretion is the enemy of order, standardization and quality.

*MARKETING FOR BUSINESS GROWTH*
THEODORE LEVITT

It's common for Contractors to be asked to provide estimates. According to *Webster's 10th Collegiate Dictionary,* estimate means "a rough or approximate calculation."

Do you want to do business with someone who gives you a rough approximation? Can you imagine basing a business decision on a rough approximation?

Well, that's what most Contractors do. It seems they've grown accustomed to using the word *estimate* without thinking about what it really means. Is it any wonder most Contractors lose money?

Enlightened Contractors, in contrast, banish the word *estimate* from their vocabulary. When it comes to estimating, just say no!

Let's say you're on a call to Mrs. Jones's house. You've surveyed the job, after which Mrs. Jones asks for an estimate.

You smile and say confidently, "We don't make estimates, Mrs. Jones. We make *promises*."

"You make *what?*" Mrs. Jones exclaims, mouth agape. It's as though you just told her you'd do the job for free.

"Promises," you reply nonchalantly. Because you know the truth about what you do: Your superb system allows you to make promises because . . . you can keep them.

You now have Mrs. Jones's rapt attention.

"Mrs. Jones, we stopped giving estimates a long time ago. At Quality Contractors, we realize that an estimate is a rough approximation of cost and time. We feel that if we can't be more exact than that, we don't deserve to be in business.

"So, Mrs. Jones, we no longer provide estimates. Now we define exactly what it will take to do the job, how much it will cost, and how long it will take, and then we give you our *written promise* that what we have agreed upon is *exactly* what we will deliver. On price, on time, and with precisely the quality you would expect from the very best Contractor in the county."

"But how can you do that?" Mrs. Jones asks, just now regaining all muscular function.

You smile. "That's what we do, Mrs. Jones. But I'll tell you this: We can do it because we did our homework."

\*    \*    \*

"But you can never be exact," Contractors have told me for years. "Close, maybe. But never exact."

I have a simple answer to that: *You have to be!* You simply can't afford to be inexact. You can't accept inexactness in yourself or in your business.

You can't go to work every day believing that your business, the work you do, the commitments you make, are all too complex and unpredictable to be exact. With a mindset like that, you're doomed to run a sloppy ship. A ship that will eventually sink and suck you down with it.

This is so easy to avoid. Because sloppiness—in thought and action—is the root cause of your frustrations.

The solution to those frustrations is clarity. Clarity gives you the ability to set a clear direction, which provides the momentum you need to grow your business.

Clarity, direction, momentum—all come from insisting on exactness.

But how do you create exactness in a hopelessly inexact world? The answer is, *You discover the exactness in your business by refusing to do any work that can't be controlled exactly.*

The only other option is to analyze the market, determine where the opportunities are, and then organize your business to be the *exact provider* of the services you've chosen to offer.

Two choices, and only two choices: (1) Look at the business you are currently doing and then limit yourself to those jobs you know you can do exactly; or (2) Start all over by analyzing the market, identifying the key opportunities in that market, and building a business that operates exactly.

What you cannot do, what you must refuse to do from this day forward, is to allow yourself to operate with an inexact mindset. It will lead you to ruin.

Which leads us inexorably back to the word I have been using throughout this book: *systems.*

Who makes estimates? Only Contractors who are unclear about exactly how to do the job in question. Contractors whose experience has persuaded them that if something can go wrong, it will—and to them!

I'm not suggesting that a Systems Solution will guarantee that you always perform exactly as promised. But I am saying that a Systems Solution will faithfully alert you when you're going off track, and will do it before you have to pay the price for it.

In short, with a Systems Solution in place, your need to estimate will be a thing of the past, both because you have organized your business to anticipate mistakes, and because you have put into place the system to do something about those mistakes before they blow up.

There's this too: To make a Promise you intend to keep places a burden on you and your managers to dig deeply into how you intend to keep it. Such a burden will transform your intentions and increase your attention to detail.

With the Promise will come dedication. With dedication will come integrity. With integrity will come consistency. With consistency will come results you can count on. And results you can count on mean that you get exactly what you hoped for at the outset of your business: the true pride of ownership that every Contractor should experience.

# CHAPTER NINE

# On the Subject of Customers

If you don't know who they are, you can't find them.

ANONYMOUS CONTRACTOR

When it comes to Contractors, the best definition of Customers I've ever heard is this:

**CUSTOMERS:** very special people who drive most Contractors crazy

Does that work for you?

After all, it's the rare customer who shows any appreciation for what a Contractor has to go through to do the job as promised. And don't Customers always think the price is too high? And don't they focus on problems, broken promises, and the mistakes you make, rather than all the times you bend over backward to give them what they say they want?

Do you ever hear these complaints? More to the point, have you ever voiced them, even to yourself?

Well, you're not alone. I have yet to meet a Contractor who doesn't suffer from a strong case of Customer confusion.

Customer confusion is about:

1. **What your Customer really wants**

2. **How to communicate with your Customer effectively**

3. **How to get your Customer to buy**

4. **How to keep your Customer truly happy**

5. **How to deal with your Customer's dissatisfaction**

6. **Who to call a Customer, and who not to**

Let's take a look at each one of these confusions.

### CONFUSION 1: WHAT DOES YOUR CUSTOMER REALLY WANT?

Your Customers aren't just people; they're very specific kinds of people. Let me share with you the six categories of Customers as seen from an E-Myth marketing perspective: (1) Tactile Customers; (2) Neutral Customers; (3) Withdrawal Customers; (4) Experimental Customers; (5) Transitional Customers; and (6) Traditional Customers.

Your entire marketing strategy must be based on

which types of Customers you are dealing with. Each of the six customer types buys products and services for very different, and identifiable, reasons. And these are:

**1. Tactile Customers get their major gratification from interacting with other people.**

**2. Neutral Customers get their major gratification from interacting with inanimate objects (a computer, a car, information).**

**3. Withdrawal Customers get their major gratification from interacting with ideas (thoughts, concepts, stories).**

**4. Experimental Customers rationalize their buying decisions by perceiving that what they bought is new, revolutionary, and innovative.**

**5. Transitional Customers rationalize their buying decisions by perceiving that what they bought is dependable and reliable.**

**6. Traditional Customers rationalize their buying decisions by perceiving that what they bought is cost-effective, a good deal, and worth the money.**

In short:

**1. If your Customer is Tactile, you have to emphasize the *people* of your business.**

**2. If your Customer is Neutral, you have to emphasize the *technology* of your business.**

**3. If your Customer is a Withdrawal Customer, you have to emphasize the *idea* of your business.**

**4. If your Customer is an Experimental Customer, you have to emphasize the *uniqueness* of your business.**

**5. If your Customer is Transitional, you have to emphasize the *dependability* of your business.**

**6. If your Customer is Traditional, you have to talk about the *financial competitiveness* of your business.**

Additionally, what your Customers want is determined by who they are. Who they are is regularly demonstrated by what they do. Think about the Customers with whom you do business. Ask yourself: In which of the categories would I place them? What do they do for a living?

For example:

**If they are mechanical engineers, they are probably Neutral Customers.**

**If they are cardiologists, they are probably Tactile.**

**If they are software engineers, they are probably Experimental.**

**If they are accountants, they are probably Traditional.**

But don't take my word for it. Make your own analysis.

### CONFUSION 2: HOW TO COMMUNICATE EFFECTIVELY WITH YOUR CUSTOMER

The next step in the Customer Satisfaction Process is to decide how to magnify the characteristics of your business that are most likely to appeal to your category of Customer. That begins with what marketing people call your Positioning Strategy.

What do I mean by *positioning* your business? You position your business with words. A few well-chosen words to tell your Customers exactly what they want to hear. In marketing lingo, those words are called your USP, or Unique Selling Proposition.

For example, if you are targeting Tactile Customers (people), your USP could be: "Superior Contracting, where the feelings of people *really* count!" If you are targeting Experimental Customers (new things), your USP could be: "Superior Contracting, where living on the edge is a way of life!" In other words, when they choose to do business with your company, they can count on your job being unique, original, on the cutting edge.

Do you get it? Do you see how the ordinary things

most Contractors do to get Customers can be done in a significantly more effective way? Once you understand the essential principles of marketing The E-Myth Way, the strategies by which you attract customers can make an enormous difference in your market share.

When applied to your business, your Positioning Strategy becomes the foundation of what we at E-Myth call your Lead Generation System.

## CONFUSION 3: HOW TO GET YOUR CUSTOMER TO BUY

If your Marketing Process begins with the identification of who your Customer is, followed by your Positioning Strategy to attract that Customer to your door, this Benchmark calls for the creation of an organized Selling Process. It's what we at E-Myth call your Lead Conversion Process. This is the system through which you consistently assure your customers that, indeed, your business was created just for them.

And just like your USP, your Lead Conversion Process demands that you organize just the right words in just the right order, to support your Customers' need for congruity. In this case, congruity means consistency—meaning that the script your salespeople use is congruent with the promise your USP made.

*And it's done in exactly the same way each time.*

While your USP makes a promise to your Customers, your Lead Conversion System helps your prospective Customers understand exactly how you intend to keep that promise when they buy your services.

## CONFUSION 4: HOW TO KEEP YOUR CUSTOMER HAPPY

Let's say you've overcome the first three confusions—now how do you keep your Customer happy? Very simple . . . just *keep* your promise! And make sure your Customer *knows* you kept your promise *every step along the way.*

In short, giving your Customers what they think they want is the key to keeping your Customers (or anyone else, for that matter) really happy.

If your Customers need to interact with people (high touch, Tactile), make certain that they do.

If they need to interact with things (high-tech, Neutral), make certain that they do.

If they need to interact with ideas (in their head, Withdrawal), make certain that they do.

And so forth.

At E-Myth, we call this your Client Fulfillment System. It's the step-by-step process by which you do the job you've contracted to do, and deliver the product you've promised.

But what happens when your Customers are *not* happy? What happens when you've done everything I've mentioned here, and they are still dissatisfied?

## CONFUSION 5: HOW TO DEAL WITH CUSTOMER DISSATISFACTION

If you have hit each step to this point, customer dissatisfaction will be rare. But dissatisfactions will happen. Here's what to do about them:

**1. Always listen to what your Customers are saying. And never interrupt while they're saying it!**

**2. After you're sure you've heard all of your Customer's complaint, make absolutely certain you understand what he or she said. You could ask, "Can I repeat what you've just told me, Mrs. Jones, to make absolutely certain I understand you?"**

**3. Secure your Customer's acknowledgment that you have heard his or her complaint accurately.**

**4. Apologize for whatever your Customer thinks you did that dissatisfied him or her even if you didn't do it!**

**5. After your Customer has acknowledged your apology, ask exactly what would make him or her happy.**

**6. Repeat what your Customer told you would make him or her happy, and get his or her acknowledgment that you heard it correctly.**

### 7. If at all possible, give your Customer exactly what he or she asked for!

But what if your Customer wants something completely unreasonable? If you've followed my recommendations to the letter, what your Customer asks will seldom seem unreasonable. That's assuming you've got the right Customer.

### CONFUSION 6: WHO TO CALL A CUSTOMER

At this stage, it's important to ask some questions:

- Which types of Customers would you most like to do business with?

- Where do you see your real market opportunities?

- Who would you like to work with, provide service for, and position your business for?

A Tactile Customer for whom people is most important?

A Neutral Customer for whom the mechanics of how you do business is most important?

An Experimental Customer for whom cutting-edge innovation is important?

A Traditional Customer for whom low cost and certainty of delivery are absolutely essential?

In short, *it's all up to you.* No mystery. No magic.

Just a systematic process for shaping your business's future. But you must have the passion to pursue the process. And you must be absolutely clear about every aspect of it.

Until you know your Customers as well as you know yourself.

Until all your complaints about Customers are a thing of the past.

Until you accept the undeniable fact that Customer Acquisition and Customer Satisfaction are more science than art.

But unless you're willing to grow your business, you better not follow any of the above recommendations. Because it will definitely grow.

# CHAPTER TEN

# On the Subject of Growth

The "rest of the world" does not sit idly "out there."
It is a sparkling realm of continual creation,
transformation, and annihilation.

*THE DANCING WU LI MASTERS*
GARY ZUKAV

Now that you've read Chapter 9 on Customers, you know what to do to acquire more of them. As many as you could want. Which brings us to the subject of growth.

The rule of business growth tells us that every business, like every child, is destined to grow! Needs to grow. Is determined to grow.

Once you've created your business, once you've shaped the idea of it, the most natural thing for it to do is to . . . *grow*! And if you stop it from growing, it will die.

Once Contractors have started their business, their most important job is to help their business grow. To nurture it and support it in every way. To infuse it with:

- Purpose

- Passion

- Will

- Belief

- Personality

- Method

As your business grows, it naturally changes. And as it changes from a little business to something much bigger, you will begin to feel out of control. And that's because you *are* out of control!

Your business has exceeded your know-how, sprinted right past you, and now it's taunting you to keep up. That leaves you two choices: Grow as big as your business demands you grow, or try to hold your business at its present level. At the level you feel most comfortable.

The sad fact is that most Contractors do the latter. They try to keep their business small, securely within their Comfort Zone. Doing what they know how to do, what they feel most comfortable doing. It's called playing it safe.

But as the business grows, so does the number of tasks, along with the scale of the tasks and their sheer size and complexity, until they threaten to overwhelm the Contractor. More people are needed. More complex tools. More money. Everything seems to be happening at the same time. A hundred balls are in the air at once.

As I've said throughout this book, *most Contractors are not entrepreneurs, are not truly business people, but technicians suffering from an entrepreneurial seizure.* So their philosophy of coping with the workload can be summarized as "just do it," rather than figuring out how to get it done through other people using innovative systems to produce consistent results.

Given most Contractors' inclination to be the master juggler in their business, it's not surprising that, as complexity increases, as work expands beyond their ability to do it, as money becomes more elusive, they are just holding on, desperately juggling more and more balls, until most collapse under the strain.

You can't expect your business to stand still. You can't expect your business to stay small. A business that stays small and depends upon you to do everything isn't a business—it's a job!

Yes, just like your children, your business must be allowed to grow, to flourish, to change, to become more. In this way, it will match your Vision.

It's either going to grow, or it's going to die. The choice is yours, but it is a choice that must be made.

The only question is, will you be ready?

# CHAPTER ELEVEN

# On the Subject of Change

Generally speaking, the way of the warrior
is resolute acceptance of death.

*THE BOOK OF FIVE RINGS*
MIYAMOTO MUSASHI

So your business is growing, which means, of course, that it's also changing. And that means it's driving you and everyone in your life crazy.

Because, to most people, change is a diabolical thing. Tell most people they've got to change, and they will crawl into a shell. Nothing threatens their existence more than change. Nothing cements their resistance more than change. Nothing.

Yet for the past 30 years, that's exactly what I've been proposing to small business owners: the need to change. Not for the sake of change, but for the sake of their lives.

I'm talking about small business owners whose

hopes weren't being realized through their business; whose lives were consumed by work; who slaved increasingly longer hours for decreasing pay; whose dissatisfaction grew as their enjoyment shriveled; whose business had become the worst job in the world; whose money was out of control; whose employees were a source of never-ending hassles, just like their Customers, their bank, and, increasingly, even their family.

More and more, these Contractors spent their time alone, in dread of the unknown, anxious about the future. And even when they were with people, they didn't know how to relax. Their mind was always on the job. They were distracted by work, by the thought of work. By the fear of falling behind.

And yet, if confronted with their condition and offered an alternative, most of these same Contractors would strenuously resist. They assumed that if there were a better way to do business, they already would have figured it out. Besides, they derived comfort in knowing what they believed they already knew, such as the limitations of being a Contractor; or the truth about people; or the limitations of what they could expect from their customers, their Sub-Contractors, their banker, their family, and their friends.

In short, most Contractors I've met over the years would rather live with the frustrations they already have rather than risk enduring new frustrations.

Isn't that true of most people you know? Rather than opening up to the infinite number of possibilities life offers, they prefer to shut their life down to respectable limits? And, after all, isn't that the most reasonable way to live?

I think not. I think we must learn to let go. I think that if we fail to embrace change, it will inevitably destroy us.

Let me share with you an original way to think about change, about life, about who we are and what we do. About the stunningly original notion of expansion and contraction.

"Our salvation," a wise man once said, "is to allow." That is, to be open, to let go of our beliefs.

That man's name was Thaddeus Golas, the author of a small, powerful book entitled *The Lazy Man's Guide to Enlightenment* (Seed Center, 1971).

Here's one of his many compelling ideas:

*The basic function of each being is expanding and contracting. Expanded beings are permeative; contracted beings are dense and impermeative. Therefore each of us, alone or in combination, may appear as space, energy, or mass, depending on the ratio of expansion to contraction chosen, and what kind of vibrations each of us expresses by alternating expansion and contraction. Each being controls his own vibrations.*

In other words, Golas tells us that the entire mystery of life can be summed up in two words: *expansion* and *contraction*. He goes on to say:

*We experience expansion as awareness, compre-hension, understanding, or whatever we wish to call it.*

*When we are completely expanded, we have a feeling of total awareness, of being one with all life.*

*At that level we have no resistance to any vibrations or interactions with other beings. It is timeless bliss, with unlimited choice of con-sciousness, perception, and feeling.*

*On the other hand, when a (human) being is totally contracted, he is a mass particle, com-pletely imploded.*

*To the degree that he is contracted, a being is unable to be in the same space with others, so contraction is felt as fear, pain, unconsciousness, ignorance, hatred, evil, and a whole host of strange feelings.*

*At an extreme (of contraction), a human being has the feeling of being completely insane, of resisting everyone and everything, of being unable to choose the content of his consciousness.*

*Of course, these are just the feelings appro-priate to mass vibration levels, and he can get out of them at any time by expanding, by letting go of all resistance to what he thinks, sees, or feels.*

Stay with me here. Because what Golas says is pro-foundly important. When you're feeling oppressed, overwhelmed, exhausted by more than you can control—contracted, as Golas puts it—you can change your state to one of expansion.

According to Golas, the more contracted we are, the more threatened by change; the more expanded we are, the more open to change.

In our most enlightened—that is, open—state, change is as welcome as nonchange. Everything is perceived as a part of ourselves. There is no inside or outside. Everything is one thing. Our sense of being an isolated individual in the world is completely transformed to a feeling of ease, of light, of joyful relationship with everything.

In fact, when we were infants, we didn't even think of change in the same way, because we lived early on in an unthreatened state. Insensitive to the threat of loss, most young children are only aware of *what is.* Change is simply another form of *what is.* Change just *is.*

When we are in our most contracted—closed—state, however, change is the most extreme threat. If the known is what I have, then the unknown must be what threatens to take away what I have. Change, then, is the unknown. And the unknown is fear.

Fear is (1) what protects what I have from being taken away; (2) the buffer that keeps me isolated from the unknown, disconnected from the rest of the world. Said in another way, fear is what keeps me alone. Separate. And here's the beauty of what Golas is saying: With this new understanding of contraction and expansion, you and I can become completely attuned to where we are at all times.

If I am afraid, suspicious, skeptical, and resistant, I

am in a contracted state. If I am joyful, open, interested, and willing, I am in an open state. Just knowing this puts me on an expanded path. Always remembering this, Golas says, brings enlightenment, which opens me even more.

Such openness gives me the ability to freely access my options. And taking advantage of options is the best part of change. Just as there are infinite ways to build a house, there are infinite ways to run your business. If you believe Thaddeus Golas, your most exciting option is to be open to all of them.

Because your life is lived on a continuum between the most contracted and most expanded—the most shut down and most open—states, change is best understood as the movement from one toward the other, and back again.

Most of the small business owners I've met see change as a thing-in-itself, as something that just happens to them. Most experience change as a threat. Whenever change shows up at the door, they quickly slam it. Many bolt the door and pile up the furniture. Some even run for their gun.

Few of them understand that change isn't a thing-in-itself, but rather the manifestation of many things. You might call it the revelation of all possibilities.

Change is where opportunity lives. Without change we would stay exactly as we are. The universe would be frozen still. Time would end.

Where we are at any given moment is somewhere

on the path between a contracted and expanded state. Most of us are in the middle of the journey, neither totally closed nor totally open. According to Golas, change is our movement from our place in the middle toward one of the two ends.

Do you want to move toward contraction, toward enlightenment? Then you must embrace change, for without it you are hopelessly stuck with what you've got.

Without change:

- We have no hope

- We cannot know true joy

- We will not get better

- We will continue to focus exclusively on what we have and the threat of losing it

All of this negativity contracts us even more, until, at the extreme closed end of the scale, we become a black hole so dense that no light gets in or out.

Sadly, the harder we try to hold on to what we've got, the less able we are to do so. So we try still harder, which eventually drags us even deeper into the black hole of contraction.

Are you like that? Do you know anybody who is?

Think of change as the movement between where we are and where we're not. There can be only two directions for change: either slipping backward or moving forward. We either become more contracted or more expanded.

The next step is to link change to how we feel: If we feel afraid, change is dragging us backward. If we feel open, change is pushing us forward.

Change is not a thing-in-itself, but a movement of our consciousness. And by being aware, we get clues to the state of our being.

Change, then, is not an outcome or something to be acquired. Change is a shift of our consciousness, of our ᵇbeing, of our humanity, of our attention, of our relationship with all other beings in the universe.

We are either "more in relationship" or "less in relationship." Change is the movement in either of those directions. The exciting part is that *we possess the ability to decide which way we go . . . and to know in the moment which way we're moving.*

Closed, open. . . . Open, closed. Two directions in the universe. The choice is yours.

What an extraordinary way to live!


There is a profound opportunity here for each of us.

Enlightenment is not reserved for the saintly. Rather, it comes to us as we become more sensitive to ourselves. Eventually, we become our own guides. Then our being says to each of us, "Open. Closed. Open. Closed."

Listen to your inner voice, your ally, and feel what it's like to be open and closed. Experience the instant of choice in both directions.

You will feel the awareness growing. It may be only a flash at first, so be alert. This feeling is accessible, but only if we avoid total contraction.

Are you totally contracted? It's doubtful. The fact that you're still reading this book suggests that you're moving in the opposite direction.

You're more like a running back seeking the open field. You can see opportunity gleaming in the distance. In the open direction.

Understand, I'm not saying that change itself is a point on the path; rather, it's the all-important movement.

Change is *in you,* not *out there.*

What path are you on? The path of Liberation or the path of Crystallization?

As we know, change can be for the better or for the worse.

If change is happening not *outside* of you, but *inside* of you, then it is for the worse only if you aren't open to it. What matters is your attitude—your acceptance or rejection of change. For change can only be for the better if you accept it. And it will certainly be worse if you don't.

Remember, change is nothing in itself. Without you, change doesn't exist! Change is happening inside of each of us, giving us clues to where we are at any point in time. Rejoice in change, for it's a sign you're alive.

Are we open? Are we closed? If we're open, good things are bound to happen. If we're closed, things will only get worse.

According to Golas, it's as simple as that. Whatever happens is where you and I are. *How* we are is *where* we are. It cannot be any other way.

For change is life.

The growth of your business, then, is its change. Your role is to go with it, to be with it, to share the joy, embrace the opportunities, meet the challenges, learn the lessons.

## THE BIG CHANGE

If all this is going to mean anything, you have to know when you're going to leave the business. At what point, in your company's rise from where it is now to where it can ultimately grow, are you going to sell it? Because if you don't have a clear picture of when you want out, your business is the master of your destiny, not the reverse.

As we stated in Chapter 2, the most valuable form of money is Equity. Unless your business vision includes your Equity and how you will use it to your advantage, you will be forever consumed by your business.

Your business is potentially the best friend you ever had. It is your business's nature to serve you, so let it. If, however, you are not a wise steward, if you do not tell your business what you expect from it, it will run rampant, abuse you, use you, and confuse you.

Change. Growth. Equity.

Focus on the point in the future when you will take leave of your business. Now reconsider your goals in that context. Be specific. Skipping this step is like tiptoeing through earthquake country.

Who knows where the fault lies waiting?

# CHAPTER TWELVE

# On the Subject of Time

We must not forget that it is not a thing that
lends significance to a moment, it is the moment
that lends significance to things.

*THE SABBATH*
ABRAHAM JOSHUA HESCHEL

I 'm running out of time!" Contractors often lament.
"I've got to learn how to manage my time more
carefully!"

Of course, deep down they know they have no
answers to these dilemmas. They're just worrying the
subject to death. Singing the Contractor's Blues.

Maybe they go to time management classes; or they
try to record where they need to be and what they need
to do during every hour of the day. Starting out before
the sun comes up and working late into the evening.

But it's hopeless. Even when Contractors work
harder, even when they keep a precise record of their
time, there's still not enough of it. It's as if they're

looking at a square clock in a round universe. Something doesn't fit. The result: They're constantly chasing the job, money, life.

And the reason is simple. Contractors don't see time for what it really is. They think of time with a small t, rather than Time with a capital T.

Yet Time is simply another word for *your life.* It's your ultimate asset, your gift at birth—and you can spend it any way you want. Do you know how you want to spend it? Do you have a plan?

At this precise moment, each of us is exactly where we are in relationship to the beginning of our Time (our Birth) and the end of our Time (our Death).

Scary, isn't it? No wonder everyone is fretting about Time. What they're really terrified of is that they're using up their life and they can't stop it.

They're accelerating toward the end with nothing to break their free fall. Their time is out of control! Understandably, this is horrifying, mostly because what they're really dealing with is death with a capital D.

They are trying to read a square clock in a round universe, trying to put time in a different perspective. All the while pretending they can manage it!

They talk about time as though it were something other than what it is. "Time is money," they announce as though that explains it.

But what every Contractor should know is that Time is life.

And Time ends! Life ends!

So now you must face the real problem, the ultimate problem! The big, walloping, unresolvable problem: We don't know how much Time we have left!

Do you feel the fear? Do you want to get over it? Then let's look at this subject called Time more seriously.

The first thing you have to do is put Time into perspective. You've got to think of it as Time with a capital T. And you've got to ask the Big Question with a capital Q.

That Big Question is: *How do I wish to spend the rest of my Time?*

Because I can assure you that if we don't ask that Big Question with a capital Q, we'll forever be asking it with a small q. We'll forever be reducing the whole of our life to the littlest things in it—to *this time,* and the *next time,* and the *last time,* all the while wondering, *What time is it?*

It's like running around the deck of a sinking ship worrying about where you left the keys to your cabin.

While doing job 5, you're worrying about the materials that haven't been delivered for job 2, or the budget for job 4, or the framer who didn't show up on job 6.

While going to sleep at 10 P.M. you're thinking about waking up at 6 A.M., worrying what you need to get done by 8 A.M. so that you can go to lunch by noon, because Murray will be there at 2 P.M. to plan what you two will do at 4 P.M. the next day.

In short, until you've answered the Big Question with a capital Q, the little questions will drive you crazy.

You must accept that you have only so much Time, that you're using up that Time second by precious second. And that Time, your Time, your life, is the most valuable asset you have. Of course, you can use your Time any way you want. But unless you choose to use it as richly, as rewardingly, as excitingly, as intelligently, as fully, as *intentionally* as possible, you'll squander it and fail to appreciate it.

Indeed, if you are oblivious to the value of Time—your Time—you'll commit the single greatest sin we can commit; you will live your life unconscious of it passing you by!

Until you deal with Time with a Capital T, you'll worry to death about time with a little t. Until you have no Time—or life—left. Then your Time will be history—along with your life.

Scary, isn't it?

Remember when we all asked, "What do I want to be when I grow up?" It was one of our biggest concerns as children.

But consider that the question doesn't include, "What do I want to *do* when I grow up?" It asks, "What do I want to *be* when I grow up?"

Shakespeare wrote: "To be, or not to be." Not, "To do, or not to do."

But when you grow up, people are always asking you, "What do you *do*?" How did the question change from *being* to *doing*? How did we miss the important distinction between the two?

How did we fail to see that even as little children, we understood the distinction between the two? After all, *being* is qualitatively different than *doing*. The real question we were asking ourselves is not what we would end up *doing* when we grew up, but who we would be.

We're talking about a *life* choice, not a *work* choice; a choice of how we spend our Time, not what we do *in* time.

Look to children for guidance. I believe that as children we instinctively saw Time as life, and tried to use it wisely. As children we wanted to make a life choice, not a work choice. As children, we didn't know—or care—that job 6 had to get done on time, on budget.

Until you see Time for what it really is—your life span—you will always ask the wrong question.

Until you embrace the whole of your Time, and shape it accordingly, you will never be able to fully appreciate the moment. Until you fully appreciate every second that composes Time, you will never be sufficiently motivated to live those seconds fully.

To be in the moment! To be in the Zone!

Until we're sufficiently motivated to live those seconds fully, we will never have cause to change the way we are. And we will never take the quality and sanctity of Time seriously.

Unless we take the sanctity of Time seriously, we will continue to struggle to catch up with something behind us. We will forever be trying to snatch the second that just whisked by.

Because if you're constantly fretting about Time, you're going to end up missing the point. The real truth about Time is this: You can't manage it; you never could. You can only *live* it.

And so that leaves you with only these questions: How do I live my life? How can I be here now, in this moment? How do I to give significance to it?

# CHAPTER THIRTEEN

# On the Subject of Work

Labor without dignity is the cause of misery;

rest without spirit the source of depravity.

*THE SABBATH*
ABRAHAM JOSHUA HESCHEL

We've come full circle to the subject of work. More than any other, this subject is the cause of obsessive-compulsive behavior by Contractors.

Work. You've got to do it every single day.

Work. If you fall behind, you'll pay for it.

Work. There's either too much or not enough.

So many Contractors describe it as "what I do when I'm busy"!

Some Contractors discriminate between the work *I could be doing* as a Contractor and the work *I should be doing* as a Contractor. But according to The E-Myth, they're exactly the same thing. The work you *could* do and the work you *should* do as a Contractor are identical. Let me describe them to you.

Contractors can do only two kinds of work: Strategic Work and Tactical Work.

Tactical Work is easier to understand, because it's what almost every Contractor does almost every minute of every hour of every day. It's called getting the job done. It's called doing business.

Tactical Work includes answering the telephone, picking up a shovel, digging a ditch, driving the truck from job to job, completing an estimate, going to the bank, and meeting with Customers.

The E-Myth says that for a contracting business Tactical Work is best defined as all the work Contractors find themselves doing to avoid doing the Strategic Work of their business.

"I'm too busy," most Contractors will tell you.

"How come nothing goes right unless I do it myself?" most Contractors complain in frustration.

Contractors say these things when they're up to their ears in Tactical Work. But what most Contractors don't understand is that if they had done more Strategic Work, they would have less Tactical Work to do.

A Contractor is doing Strategic Work when he or she asks the following questions:

- Why am I a Contractor?

- What will my business look like when it's done?

- What must my business look, act, and feel like in order for it to compete successfully?

- What are the Key Indicators of my business?

Please note that I said a Contractor *asks* these questions when he or she is doing Strategic Work. I didn't say these are the questions a Contractor necessarily *answers.*

And that is the fundamental difference between Strategic Work and Tactical Work. Tactical Work is all about *answers:* How to do this. How to do that.

Strategic Work, in contrast, is all about *questions:* What? Why? Who? When? Where? What business are we really in? Why are we in that business? Who specifically is our business determined to serve? When will I sell this business? Where will this business be doing business when I sell it? And so forth.

Not that Strategic Questions don't have answers. Contractors who commonly ask Strategic Questions know that once such a question has been asked, they're already on their way to envisioning the answer. Within the Strategic Question resides the answer. Question and answer are part of a whole. You can't find the right answer until you've asked the right question.

Tactical Work is much easier, because the question is always more obvious. In fact, you don't ask the Tactical Question; instead, the question arises from a result you need to get or a problem you need to solve. Digging a ditch is Tactical Work. Framing a house is Tactical Work. Fixing a leak is Tactical Work. Firing a Sub-Contractor is Tactical Work.

Tactical Work is the stuff you do every day in your business. Strategic Work is the stuff you plan to do to create an exceptional company.

In Tactical Work, the question comes from *out there,* rather than *in here.*

The Tactical Question is about something *outside* of you, whereas the Strategic Question is about something *inside* of you. The Tactical Question is about something you *need* to do, whereas the Strategic Question is about something you *want* to do. Want versus Need.

If Tactical Work consumes you:

- You are always reacting to something outside of you

- Your business runs you; you don't run it

- The job runs you; you don't run it

- Your Sub-Contractors run you; you don't run them

- Your life runs you; you don't run your life

You must understand that the more Strategic Work you do, the more intentional your business, your jobs, and your life become. *Intention* is the byword of Strategic Work.

Everything on the outside begins to serve you, to serve your Vision, rather than forcing you to serve it. Everything you *need* to do begins to respond to everything you *want* to do. It means you have a Vision, an Aim, a Purpose, a Strategy, an *envisioned* result.

Strategic Work is the work you do to *design* your business, to design your life.

Tactical Work is the work you do to *implement* the design created by Strategic Work.

Without Strategic Work, there is no design. Without Strategic Work, all that's left is keeping busy.

Let's look at someone who learned the important difference between Strategic Work and Tactical Work. Let's look at the story of Three Day Kitchens.

# CHAPTER FOURTEEN

# The Story of Three Day Kitchens

Think like a man of action; act like a man of thought.

HENRI BERGSON

Now let's meet someone who is the embodiment of the principles we've discussed here. His name is Marino Santos; his company is Santos Construction.

Marino Santos was a framer; he and his small crew Sub-Contracted the framing of homes in Southern California, Arizona, Nevada, Colorado, wherever the work took them.

To the general contractors who hired them, they were simply known as Santos's crew. Within Santos Construction—more like a small band of men than an actual business—they thought of themselves as *los apasionados sin igual*—"without equal."

Everyone in the trade knew Santos's crew. They

were the stuff of folklore. No one disputed that they were the best, but it was more than that. A mystique surrounded them wherever they went.

They were a tight bunch. At breaks they hung together, drinking coffee, eating burritos, and whispering among themselves—a cluster of stars.

But when Marino Santos and his crew went to work, there was nothing quiet about them. Their framing hammers fairly flew. Walls went up in record time, first one house and then another. The sounds of their tools at work reverberated off the hills.

Often they worked to music. For Santos's crew, every day on the job was a performance, a dance, a crusade. *Los apasionados sin igual.* It was what they did, yes, but more. It was what they lived to do. It was who they were.

And then one July morning, on his way to a job in Barstow, California, Marino Santos's pickup blew a tire and left the road going 90 miles an hour. It turned over five times, finally coming to rest upside down against a boulder.

Marino Santos lay trapped in his truck for 13 hours before help arrived. His back was broken. What a miracle that he survived the accident, everyone said. He failed to see the miracle. In fact, he was blind to everything except one unassailable fact: He was finished with the framing business.

What does a framer do when he can't frame anymore? Especially one as driven to excellence as Marino Santos? This framer drank, and for the next six months was rarely sober. Sometimes he railed at the night, flinging his empty whiskey bottle through the closed bedroom window and into the street. For hours, he would sit in his wheelchair amid the shards of broken glass, screaming at the injustice.

His crew came to visit him every day. They cried with him. They sat silently with him. They played music and drank with him. And for a long time, nothing changed. Then one day, it did. Marino Santos was getting better.

At last, Santos called his crew together and apologized for his stupidity. "I don't want to be stupid anymore," he told them. "It's time to start a new business."

The new business would be in construction, but in what field? Yes, that would take some study. In the meantime, he knew this: His new business would have as great an impact on the people around him as the old business had had.

One day, Santos addressed his crew. "I have thought about this a good deal," he said. "It comes down to this: Either we work for a living, like burros, until we can't work anymore. Or we find a way to build a business that works for us. We think about this business, we put our minds to it, and we shape it so that it works like we have learned to work, with precision, joy, and energy.

"But we must build it so that it works even without people like us. We must make it easy for people who are not like us to act like us. We must learn how to give our fierce pride to people who do not possess it naturally. And we must make it possible for everyone who works in our business to become as good as we are. That will be our gift to them."

Santos paused and looked soberly into his men's eyes. "Brothers, what I'm suggesting is a risky venture. I have no way of knowing if it will be successful or not. But I know in my heart it is the path for us to follow."

Santos instructed his men to take jobs in a segment of the construction industry that was new to them. They sorted the industry into new construction and reconstruction, commercial and residential. They then broke those parts down into subparts, which they analyzed using certain criteria.

The new business had to be in a segment of the industry that (1) had consistent growth; (2) did not rise and fall dramatically with the economy; (3) essentially repeated the same tasks from job to job; (4) wasn't capital intensive to start or maintain; (5) could be operated independently of other Contractors—that is, they could secure, start, and complete a contract without having to depend on general Contractors or other Sub-Contractors to do their parts of the job.

Every night the men gathered in Santos's kitchen to have a cold beer and report their findings. Often they argued. But gradually, as the men became smarter about their mission and more eloquent in their expression of it, the arguments became less heated.

Santos's strategy was simple: As one specialty after another was eliminated as an option, those employed in that field left their jobs and found work in one of the remaining fields, until, in the end, everyone was working in the same field, the one of choice.

And that's exactly what happened. After 2½ years of dedicated work, research, and planning, all signs pointed to the kitchen remodeling business. This was the birth of Three Day Kitchens.

How many people do you know who are willing to devote such effort, intelligence, care, and attention to choosing the right path. Most of us just stumble ahead, hoping it will turn out all right in the end.

Even with all that research, Santos still wasn't satisfied. He and his crew installed hundreds of kitchens—just for practice. Every imaginable problem was confronted, dealt with, and overcome. It would be another 2 years before the crew took on their first kitchen for pay.

At night they gathered to discuss the problems they had faced that day on the job, analyzing every peculiarity, every exception. No one, they were sure, had ever devoted as much time and intelligence to solving kitchen remodeling problems. They were determined to get it right.

Their goals were (1) to create a kitchen remodeling system that could produce an absolutely predictable result in the hands of novice workers trained only in their system; and (2) to figure out how to renovate or remodel any kitchen within 3 days, at a cost lower than the competition.

Actually, Santos and his crew discovered that the competition was not a problem. The crew reported daily on the waste, inefficiency, apathy, and lack of management on the jobs they were working. No, other contractors would not determine the fate of Three Day Kitchens. Marino Santos and his men would.

They studied every variation on every theme, created a variety of preplanned kitchen solutions to address each one of them, and devised a preprogrammed construction and installation strategy for each type of kitchen. Then they recruited, hired, and rigorously trained a small crew of inexperienced technicians in their construction and installation system. Finally, they created a management system to ensure that their system would be used exactly as planned, every time.

Then they practiced . . . practiced . . . practiced . . .

In the end, they invented a kitchen remodeling system that allowed them to install a kitchen in 3 days or less—guaranteed!


After all those years of preparation and study, Santos's first paid kitchen was a sweet moment for him and his crew. It went off without a hitch, just as they knew it would. They had planned it that way and practiced until nothing was taken for granted.

Their first Customer was appropriately astonished. Not only was the job done exactly as promised, but the workers were clean, well organized, and fastidious— "almost joyful," the Customer enthused.

"How do you find such good people?" she asked.

Marino Santos smiled. "I wish I knew," he said.

# CHAPTER FIFTEEN

# On the Subject of Taking Action

> Right Action cannot take place without
> Right Knowing . . . and neither can our knowing be called
> "right" if it does not produce Right Action.
>
> *THE PATH OF ACTION*
> JACK SCHWARZ

You should now be clear about the need to organize your thoughts first, then your business. Because the organization of your thoughts is the foundation for the organization of your business.

If we try to organize our business without organizing our thoughts, we will fail to attack the problem.

We have seen that organization is not, first and foremost, Time Management. Nor is it People Management; nor is it tidying up desks or alphabetizing files. No, organization is first, last, and always cleaning up the mess of our minds.

By learning how to *think* about the business of Contracting, by learning how to *think* about our priorities, by learning how to *think* about our lives, we prepare ourselves to do righteous battle with the forces of failure.

Right Thinking leads to Right Action—and now is the time to take action! Because it is only through action that we can translate thoughts into movement in the real world. Our being will take form as an action in the real world, and, in the process, find fulfillment.

So, first we *think* about what we want to do. Then, having thought it, we must *do* it. Only in this way will we be fulfilled and serve the purpose for which God brought us into the world.

How, you ask, do we do that? How do we put the principles I've just shared with you to work in our Contracting business?

To find out, accompany me down the path once more:

1. *Create a story about your business.* Remember The Ethical Home? Do you have your own version? Your story should be an idealized version of your Contracting business, a vision of what the preeminent Contractor in your field should be and why. Your story must become the very heart of your business. It must become the spirit that mobilizes it, as well as everyone who works for it, buys from it, sells to it, and lends to it. Without this story, your business will be reduced to plain work.

2. *Organize your business so that it breathes life into your story.* Unless your business can

faithfully replicate your story in action, it all becomes fiction. In that case, you'd be better off not telling your story at all. And without a story, you'd be better off leaving your business the way it is and hoping for the best.

Here are some tips for organizing your Contracting business:

- Identify the key functions of your business

- Identify the essential processes that link those functions

- Identify the results you have determined your business will produce

- Clearly state in writing how each phase will work

Take it step-by-step. Think of your business as a program, a piece of software, a system. It is a collaboration, a collection of processes dynamically interacting with one another.

Of course, your business is also people.

3. *Engage your people in the process.* Why is this the third step rather than the first? Because, contrary to the advice most business experts will give you, *you must never engage your people in the process until you yourself are clear about what you intend to do.*

The need for consensus is a disease of today's addled mind, a product of our troubled and confused times. When people don't know what to believe in, they often ask others to tell them. To ask is not to lead but to follow. *The prerequisite of sound leadership is first to know where you wish to go.* And so, "What do *I* want?" becomes the first question; not, "What do *they* want?" In a business of your own, the Vision must first be yours. To follow another's Vision is to abdicate your personal accountability, your leadership role, your true power.

In short, the role of leader cannot be delegated or shared. And without leadership, no Contractor's business will ever succeed.

Despite what you have been told, *Win-Win* is a *secondary* step, not a primary one. The opposite of *Win-Win* is not necessarily *They Lose.* Let's say "they" can win by choosing a good horse. The best choice will not be made by consensus. "Hi, guys, what horse do you think we should ride?" will always lead to endless and worthless discussions. By the time you're done jawing, the horse will have already left the post!

Before you talk to your people about what you intend to do in your business and why you intend to do it, you need to reach agreement with yourself.

Then, once you have agreement with yourself, it is critical that you engage your people in a discussion about what you intend to do and why. You do that by being clear with them and with yourself.

It's important to know (1) *exactly* what you want; (2) how you intend to proceed; (3) what's important to you and what isn't; and (4) what you want the business to be, how you want it to act, and where you want it to go.

The story is paramount, because it is your Vision. Tell it with passion and conviction. Tell it with precision. Never hurry a great story. Let it be unveiled, slowly. Don't mumble or show embarrassment. Never apologize or display false modesty. Look your audience in the eyes and tell your story as though it were the most important one they'll ever hear about business. Your business. The business into which you intend to pour your heart, your soul, your brains, your imagination, your Time, your money, and your sweaty persistence.

Get into the storytelling zone. Behave as though it means everything to you. Show no hesitation when telling your story.

These tips are important because you're going to tell your story over and over—to prospective customers, to new and old employees, to suppliers, to lenders, to your Sub-Contractors, to your family and friends. You're going to tell it at your church or synagogue; in the field; on the job; to your card-playing or fishing buddies; and to organizations such as Kiwanis, Rotary, YMCA, Hadassah, and Boy Scouts. There are few moments in your life when telling a great story about a great business is inappropriate.

You must love your story to succeed. Do you think Walt Disney loved his Disneyland story? Or Ray Kroc his McDonald's story? What about Dave Smith at Federal Express? Or Debbie Fields at Mrs. Fields cookies? Or Tom Watson Jr. at IBM?

Do you think these people loved their stories? Do you think others loved (and *still* love) to hear them? I daresay, all successful entrepreneurs have loved the story of their businesses.

Because that's what true Contractor-entrepreneurs do: *They tell stories that come to life in the form of their businesses!* They invent their own versions of The Ethical Home. And then they *live* it!

Remember, a great story never fails. A great story is always a joy to hear.

First, you need to be completely clear with your people about the story of your business. Second, you need to be clear with them about the *process* your business must go through in order for your story to become a reality.

I call this the Business Development Process. Others call it reengineering, continuous improvement, reinventing your business, or total quality management. Whatever you call it, you must take three distinct steps to succeed:

1. *Innovation*. **Continue to find better ways of doing what you do.**

2. *Quantification*. **Once that is achieved, quantify the impact of these improvements on your business.**

3. *Orchestration*. **Once these improvements are verified, orchestrate this better way of running your business so that it becomes your standard, to be repeated time and again.**

In this way, the system works—no matter who's using it. And you've built a business that works consistently, predictably, *systematically*. A business you can depend on to operate exactly as promised, every single time.

Your Vision, your People, your Process—all linked.

A superior Contracting business is a creation of your imagination, a product of your mind. So fire it up and get started!

For three decades, I've applied The E-Myth principles I've shared here to the successful development of thousands of small businesses throughout the world. Many have been Contracting businesses. Plumbers, roofers, electricians, landscapers, remodelers, every imaginable kind of Contractor.

And I must tell you that there is no greater reward than seeing these E-Myth principles put to work in the lives of so many people. Those rewards include seeing these changes:

- Lack of clarity—clarified

- Lack of organization—organized

- Lack of direction—shaped into a path, clearly, lovingly, passionately pursued

- Lack of money or money poorly managed— understood instead of coveted, created instead of chased, wisely used instead of squandered, intelligently invested instead of misspent

- Lack of committed people—transformed into people coming together as a community, working in harmony toward a common goal, discovering each other and themselves in the process, all the while expanding their understanding, their know-how, their interest, their attention

After working with so many Contractors, so many small business owners, I know that a business can be much more than what most become. I also know that nothing is preventing you from making your business all that it can be. It takes only desire and the perseverance to see it through.

To your business, and your life, good growing!

For more information about how *you* can put The E-Myth to work for you and your business, visit me at:

*www.E-Myth.com*

and click on . . .

**The E-Myth Contractor Website.**

Or call us at:

1-800-221-0266

**and ask for your**

**FREE**

**E-Myth Experience!**

# HarperBusiness

**Books by Michael E. Gerber:**

## THE E-MYTH REVISITED
*Why Most Small Businesses Don't Work and What to Do About It*
ISBN 0-88-730728-0 paperback
ISBN 0-694-51530-2 audio

A new and totally revised edition of Michael E. Gerber's 500,000-copy underground bestseller, *The E-Myth*. In order to enable you to grow your business in a predictable and productive way, Gerber:

- dispels the myths surrounding starting your own business
- shows how commonplace assumptions can get in the way of running a business
- walks you through the life of a business
- shows how to apply the lessons of franchising to ANY business
- draws the vital distinction between working on your business and working in your business

## THE E-MYTH MANAGER
*Why Management Doesn't Work—and What to Do About It*
ISBN 0-88-730959-3 paperback

Drawing on lessons learned from working with more than 15,000 small, medium-sized, and very large organizations, Gerber has unearthed the arbitrary origins of commonly held doctrines and discovered the truth behind why management doesn't work—and what to do about it.

Offering a fresh, provocative alternative to management as we know it, Gerber explores why every manager must take charge of his own life, reconcile his own personal vision with that of the organization, and develop an entrepreneurial mind-set to achieve true success.

**Available wherever books are sold, or call 1-800-331-3761 to order.**